# Instrumentation in Scientific Research

## ELECTRICAL INPUT TRANSDUCERS

# Instrumentation in Scientific Research

## ELECTRICAL INPUT TRANSDUCERS

### Kurt S. Lion

ASSOCIATE PROFESSOR

MASSACHUSETTS INSTITUTE OF TECHNOLOGY

McGRAW-HILL BOOK COMPANY, INC.

New York    Toronto    London

1959

INSTRUMENTATION IN SCIENTIFIC RESEARCH

# *Preface*

This book represents a part of a larger undertaking, the goal of which is collection, analysis, and organization of the physical methods used for experimental research or for the development of scientific or technical instruments. The purpose of this volume is to offer research workers, engineers, and students a reasonably complete collection of the existing basic methods and systems used as input transducers in electrical instrumentation. Its use as a reference book should minimize literature search and should provide information on particular methods as well as criteria by which the relative merits of existing methods and systems may be evaluated.

However, I have tried to make the book more than just a collection of unrelated facts and figures. In it, there is an attempt to establish order in the vast field of experimental methodology and instrumentation and to provide an organization which permits logical and simple grouping of the numerous varieties of methods and systems described in the literature. I feel that the establishment of such a logical organization is important, not only in order to assure adequate coverage of the material but also because instrumentation or methodology is not a bag of tricks mastered only by memorizing a huge collection of facts and gadgets, but is a unified field, a branch of physics, which can be taught like any other logical science. I do not claim credit for this crystallization process; such a development is natural, and several attempts in this direction have already been made.

A further intention has been on my mind while preparing this manuscript. I have frequently observed that creative accomplishments in the field of research methods and instrumentation are not

v

the result of expertness in mathematics only, nor of acrobatics in experimental techniques, but are often the consequence of a clear understanding of the physical principles or mechanisms underlying a process. The research worker or engineer frequently scans the field for such physical principles and creates a new solution by some kind of analogy. By furnishing the reader with a simple description of the physical mechanisms underlying each method, I hope to render him a service which will facilitate the finding of new and better methods and instruments.

The question has been raised whether a book of such general scope is justifiable or whether it would not be more reasonable to provide each science, such as the physical, the biological, or the engineering sciences, with its own book containing only those methods and systems that are of primary interest in each science. Many years of teaching research methods and instrumentation to students and research workers in different fields have convinced me that a specialized approach is of limited value, that the restriction to research methods in any one field leads to unnecessary duplication of work (the existing literature furnishes ample proof of this), and that a broad knowledge of the entire field of methodology is demanded, in particular by the experienced research worker.

Of course, it would be impossible (and useless) to describe the innumerable methods and instruments that are being used, even if one limits oneself to the electrical systems only. However, all electrical instruments are composed from _elements_, such as input transducers, amplifiers, filters, integrators, output transducers, etc. These elements recur, in one form or another, in many applications; since their number is limited, they can be adequately discussed without the danger of superficiality. They may be synthesized into complete systems in accordance with the well-known principles and laws that govern the behavior of such physical systems. I have tried to furnish enough data for the reader to make an intelligent choice in the selection and combination of the elements. The final synthesis, however, is frequently a matter of engineering art, or the art of the research worker, and cannot be taught.

The level of presentation is chosen for those who speak the language and know the concepts of elementary physics and have at least a moderate mathematical knowledge. The physics of each method has been given preference over the technical means. Mathematical derivations are omitted; where the derivation is not obvious it may be found in the references.

In the selection of the material, I have limited myself to those systems that are commonly in use and to some unorthodox methods that seemed to hold some promise for further development. Older instruments are included if the principle involved is of general interest or possibly applicable in other techniques. Experience shows that old or neglected methods are frequently revived after having undergone a metamorphosis or some improvement. Some special methods, like those used for the establishment of standards, are omitted.

Different parts of this book have been written at different times; in general, it reflects the status of the art of instrumentation between the years 1955 and 1957. However, the book is not intended to be the most "modern-minded" text, but rather a presentation of the basic methods. The development of these basic systems is slow and only moderately affected by new publications.

Information on instrumentation is scattered widely throughout the entire literature in physics, biology, medicine, and the engineering sciences. The classic book on methodology is the "Praktische Physik" originated by F. Kohlrausch (last edition edited by H. Ebert and E. Justi and published by Teubner Verlagsgesellschaft m.b.H., Stuttgart, 1956). A number of periodicals are exclusively devoted to this science, such as the *Review of Scientific Instruments* (American Institute of Physics, New York), the *Journal of Scientific Instruments* (The Institute of Physics, London), and the *Archiv für technisches Messen* (R. Oldenburg, München). A considerable contribution has been made by the collections of books published after World War II by laboratories operated by the government (e.g., M.I.T. Radiation Laboratory Series, National Nuclear Energy Series, both McGraw-Hill Book Company, Inc., New York). I have freely drawn from these sources as well as from summarizing papers and reviews. Also, I have not hesitated to include material found in pamphlets and advertising literature of commercial companies. These papers are sometimes clearer and less pretentious than "scientific" publications, and I believe that a great deal of our postgraduate education comes from such company literature. The similarity between teaching and advertising has been strikingly demonstrated by Gilbert Highet in "The Art of Teaching" (Alfred A. Knopf, Inc., New York, 1950).

I have frequently found that a very extended bibliography is of little help. Therefore, I have limited myself to those references which seemed most important. They are not always the first descriptions of any method nor the most extensive publications but generally those that seemed to be most informative or most clearly written. I am aware that by doing so I have neglected priorities, but such may be permissible in a book which is not intended as an historical source book. The reader will, in general, find the original source for any one method by following up the references. Only references in ordinary scientific and technical periodicals or textbooks are quoted. References to semipublic material, such as theses or technical reports of research-sponsoring agencies, are used only occasionally. Experience shows that a great part of such semipublic material eventually finds its way into the regular literature. Also, patents are quoted occasionally. There is a wealth of information in the patent literature, but it would require a fairly large organization to make it available in a practical form.

A problem which defies solution at the present time is that of units and dimensions. It would simplify matters considerably if a book could be written with the use of one system throughout, e.g., the mks system,[1] in particular since almost everybody concedes the superiority of a metric system. Unfortunately the physicist uses the cgs and the mks systems, the medical research worker frequently a hybrid system, the engineer, in particular the mechanical engineer, and many industries the English (inch-pound, etc.) system. I see no way out of this difficulty. The complexity of the situation is increased by a desire of certain authors to invent new, unnecessary names for dimensions ("dyne-five newton," "pharos") and, even worse, to form new derivations from such names ("pragilbert/ weber"). The situation becomes intolerable with the introduction of the "mho" (ohm$^{-1}$) which, fortunately, has not (or not yet) given rise to the "tlov" (volt$^{-1}$). I wish to state, with due humility, that the essence of any science is simplicity, and that it must be considered an asset to derive the infinite number of dimensions from a combination of the smallest possible number of basic units. However, until agreement has been reached on the question of units and dimensions, this book reflects the deplorable status of the art.

I should like to express my gratitude to the Massachusetts Institute of Technology, in particular to Dr. I. W. Sizer, head of the department of biology, who has freed me from my teaching duties for one year.

I am particularly thankful to the Office of Naval Research, which has sponsored the greater part of the research activity involved in writing this book and provided interest and help that have been a

[1] Differences exist also in the mks system. The "k" in the American mks system is the unit of mass, in the European mks system it is frequently the unit of force.

constant source of encouragement. I should like to mention specifically Mr. F. B. Isakson and Lt. R. Gomory from the Office of Naval Research, Washington, and, former members of the same office, Lt. Comdr. F. L. Thomas, Lt. D. de Graaf, and Mr. L. M. McKenzie, and, from the Boston office, Capt. J. F. Benson, Dr. C. F. Muckenhaupt, and Mr. T. B. Dowd.

Acknowledgment is also due to Mr. D. A. Berkowitz, Mr. R. H. Davis, Mr. P. Felsenthal, Dr. W. L. Harries, and Dr. G. F. Vanderschmidt, who have been working with me at different times and who have given me valuable assistance, and to Mrs. E. Duncan for her competent help in preparing and typing the manuscript.

I am aware that a book of this kind is bound to be incomplete and contain errors. I should appreciate any information that will lead to desirable corrections in the text, and I hope that with such help something of value may grow in time.

*Kurt S. Lion*

# Contents

# Introduction

The purpose of instrumentation is to obtain information about the physical or chemical nature of an investigated object or process, or to control an object or a process in accordance with such information. Information is required or supplied either in a continuously variable form, such as the deflection of a meter (analogue systems), or in discontinuous steps, such as by counting (digital systems).

In general, the simpler the means it involves, the better the instrumentation is. Many problems lend themselves to a solution by direct observation or by simple mechanical means. The possibility of such simple and direct methods should be investigated before a decision is made to use the more complex electrical methods. However, electrical methods frequently offer a number of advantages, such as high-speed operation or the possibility of simple processing of the results, e.g., amplifying, filtering, differentiating, storing, or telemetering.

All electrical instruments are composed from elements. In the sense used in this book an element is a functional unit, a building block in a block diagram, which performs one particular task or which solves one particular problem. An element is not a component part, although at times a component part can serve as an element. In the field of electrical instrumentation one can distinguish three groups of elements:

$A$. Input transducers: converting a nonelectrical quantity into an electric signal (e.g., strain gauge, photoelectric cell)

1

*B*. Modifiers: converting an electric signal into another modified electric signal (e.g., amplifier, filter)

*C*. Output transducers: converting an electric signal into a non-electrical quantity (e.g., meter, strobotron tube)

This book deals with input transducers; the standard transducer systems, or "basic solutions," are described, as well as a number of variations of each basic theme.

## General Properties of Instrumentation Elements

All instrumentation elements are characterized by the following basic criteria:

A. INPUT CHARACTERISTICS

1. *Type of Input*. This is the physical agent acting on the element and producing an output signal. In the case of an input transducer the input can be any physical quantity; in a modifier or output transducer it may be a voltage, a current, an impedance, or a variation or function of time of these quantities (e.g., time derivative, frequency, phase).

2. *Useful Range of the Input Quantity for Which the Element Can Be Used*. The lower limit of the useful input range of an element is, in general, imposed by the instrument error or by the unavoidable noise originating in the element. The upper limit of a useful input level is generally reached when excessive distortion of the signal sets in or when the input signal tends to damage the element.

3. *Effect of the Element upon the Preceding Stage or upon the Object under Investigation*. The magnitude of this effect can frequently be expressed by the *input impedance* of the element or by the amount of force, energy, or power required from the stage preceding the element.

If the physical mechanism involved in the operation of an element is such that the input quantity *actuates*, or *drives*, the element (e.g., if the input quantity is an electric signal and the element to which it is applied is a galvanometer), a maximum of power is transferred from the preceding stage to the element if the input impedance of the element matches the output impedance of the preceding stage. However, if the input quantity only *controls* the element (e.g., a voltage source applied to the control grid of a vacuum tube), impedance matching between the output from the source and the input of the element is not required.

High input impedance is desirable if the output from the preceding stage is furnished in the form of a voltage and if a minimum of current or power is to be drawn from the source. Low input impedance is preferred if the output from the preceding stage is furnished in the form of a current and if a minimum of power is to be drawn from the source, or if the presence of a large impedance in the input of the element adversely affects the signal level or the operation of the source.

The input impedance (or the input force, energy, or power) is not the only quantity which may influence the preceding stage and cause a variation of the signal. Sometimes a potential difference exists between the input terminals of an element and may have an undesirable effect on the preceding stage.

## B. TRANSFER CHARACTERISTICS

1. *Transfer Function.* This is the relationship between the magnitude of the input quantity $Q_i$ and the output quantity, or the result, $Q_o$.

$$Q_o = f(Q_i) \tag{1}$$

The differential quotient

$$\frac{dQ_o}{dQ_i} = S$$

is the *sensitivity* of the element. In general, the sensitivity varies with the input magnitude $Q_i$; in the special—although frequent— case of a linear transfer function the sensitivity is

$$\frac{dQ_o}{dQ_i} = \frac{Q_o}{Q_i} = S'$$

and is constant throughout the range of the element. In some applications the quantity $S$ is called "gain" or "attenuation factor," also "scale factor"; in others, the reciprocal value $1/S$ is called the "scale factor." The value of $S$ is sometimes expressed in logarithmic (decibel) notation.

2. *Instrument Error.* In general an instrument will not follow Eq. (1) correctly but will have an output

$$Q_o' = f(Q_i) + F \tag{2}$$

where $F$ is the (absolute) error of the result or of the output quantity, i.e., the deviation of the observed output quantity $Q_o'$ from the correct value $Q_o$.

$$F = Q_o - Q_o'$$

The error can be expressed either in terms of the output or input quantity. The fractional error is $F_f = F/Q_o$, which is approximately equal to $F/Q_o'$; the percentage error is $F_{\text{per cent}} = F/Q_o \times 100$.

The error is usually complex. It is practical to distinguish the following components:

SCALE ERROR. (1) The observed output may deviate from the correct output by an amount which is constant throughout the entire range of the instrument (additive constant, zero displacement). (2) The observed output may deviate from the correct value by a constant factor. (3) The experimentally observed transfer function may deviate from that postulated by theory (nonconformity). In particular, if a linear relationship between input and output quantity is postulated but not experimentally realized, the error is called nonlinearity, or nonlinear distortion. (4) The output may depend not only upon the applied input but also upon the past history of the element, i.e., upon the input formerly applied to the element (hysteresis error).

DYNAMIC ERROR. The output does not follow the variations with time of the input precisely or it depends upon time functions such as a time derivative or the frequency of the input quantity.

NOISE AND DRIFT. A signal originating in the element and varying with time appears at the output terminals or is superimposed on the output signal. The magnitude of this noise or drift output is, in principle, independent of the magnitude of the signal applied to the input. If information is available about the statistical nature of the noise output, it is possible sometimes to distinguish between the desired output signal and the undesirable noise.[1]

If a human observer reads the output, he is a part of the instrument system, and his reaction must be included in the consideration of errors. For instance, a scale error can be caused by parallax in scale reading, or a dynamic error can be caused by the observer's reaction time or psychological anticipation of an expected result (the "personal equation" in the observation of passage instruments in astronomy).

3. *Response to Environmental Influences.* The performance of an instrumentation element is fully described by the transfer function and the errors, as mentioned above, providing the instrument is in a constant environment and is not subjected to external disturbances. If instruments are subjected to environmental influences such as changes of temperature, pressure, or acceleration, changes of magnetic or electric fields, or changes of the supply voltages, variations

---

[1] K. S. Lion and D. F. Winter, *Electroencephalog. and Clin. Neurophysiol.*, **5**, 109 (1953).

of the transfer function and of the errors may result. Although such variations may lead to deviations in the instrument output, for practical reasons they are separated from the instrument errors described under part 2 above.

For other definitions of error terms, see, for instance, J. D. Trimmer, "Response of Physical Systems," John Wiley & Sons, Inc., New York, 1950, and C. S. Draper, W. McKay, and S. Lees, "Instrumentation Engineering," McGraw-Hill Book Company, Inc., New York, vol. 1, 1952; vol. 2, 1953; vol. 3, 1955. J. Mandel and R. D. Stieler, *J. Research Natl. Bur. Standards*, **53**, 155 (1954), and *Tech. News Bull. Natl. Bur. Standards*, **40**, 139 (1956).

## C. Output Characteristics

1. *Type of Output.* Input transducers and modifiers have an electric output which may be a voltage, a current, an impedance, or a variation or a time function of these magnitudes. The output from an output transducer is a nonelectrical quantity and is, in general, of a mechanical, thermal, or optical nature.

2. *Useful Output Level or Range.* Like the input range, the useful output range is limited at the lower end by noise considerations; an upper limit is set by the maximum useful input level. Increase of the output level is frequently technically possible (e.g., a greater gain of an amplifier or a finer suspension of a galvanometer). However, since such increases are frequently connected with an increase in noise level, they may not offer an advantage.

3. *Output Impedance.* The output impedance determines the amount of power that can be transferred from the element to the succeeding element or stage at a given output-signal level. If the output impedance of an element is low compared to the input impedance of the succeeding stage, the output has the character of a constant-voltage source. If the output impedance is high, the output has the character of a constant-current source.

The description of instrumentation elements given in this book follows the general outline of characteristics enumerated above, whenever the necessary information has been available.

The material is organized according to *input quantities;* e.g., an element that converts pressure into resistance variations will be found under Pressure Transducers (1-5). However, only such systems which convert pressure variations *directly* into electric signals will be found in this section. If the pressure is first converted into a displacement (e.g., in a pressure capsule or Bourdon tube) and the displacement is converted into a signal by an electrical system, such systems will, naturally, be found under Displacement Transducers (1-2).

The reader should be reminded that the book deals only with *electrical* transducer systems. Systems based on other physical principles, such as optical or acoustical systems, are occasionally included when it was felt that they are closely related to the field of electrical transducers.

The described breakdown of instruments into their elements not only permits a satisfactory organization of instrumentation systems, but also furnishes a logical code for the identification of instruments of any kind. Such a code can be used much in the way a chemist uses the symbolic description of chemical compounds, either for the characterization of instruments or for information retrieval in the instrumentation literature.

# 1

# *Mechanical Input Transducers*

## 1-1. Transducers for Linear Dimensions (Length or Thickness Gauges)

The following transducer systems furnish an electric signal which can serve as a measure of a linear distance between two limits, of a thickness or of a length. The systems are based on either a resistance variation (1-11a),[1] a variation of current flow pattern (1-11b), an inductance variation (1-12), or a capacitance variation (1-13). A method based upon the electric breakdown voltage is described in 1-14. Thickness of test objects can also be measured with sonic methods (1-15) and from the absorption of radiation (alpha-, beta-, gamma-, and X-ray gauges, 1-16).

Distance and thickness measurements can also be made with the help of displacement transducers (described in 1-2). Further methods, primarily applicable to liquids, are described in 1-4.

For a summarizing review, see George Keinath, The Measurement of Thickness, *Natl. Bur. Standards Circ.* 585, Jan. 20, 1958.

### 1-11. RESISTIVE SYSTEMS

*a. Contact on Both Sides.* The length of a wire or the thickness $t$ of a prismatic body with the resistivity $\rho$ clamped between two elec-

---

[1] Cross references in this book are multiple-numbered. "1-1" refers to chapter 1, part 1; "1a" designates the section of the part and its lettered subdivision.

trodes $A$ and $B$, as shown in Fig. (1-1)1, can be determined from the measurement of its resistance $R$.

$$R = \rho \frac{t}{a} \tag{1}$$

where $a$ is the cross-sectional area of the prism and of the electrodes.

The method requires the knowledge of the resistivity of the test object and is applicable for the measurement of isotropic and homo-

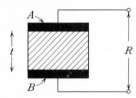

geneous test objects having a resistivity less than of the order of $10^6$ ohm-cm. If the resistivity of the test object is much higher, the resistance between the electrodes is frequently influenced by surface conduction.

If the cross section of the test object changes abruptly, Fig. (1-1)2, or if the electrodes are smaller than the test object, Fig.

FIG. (1-1)1.    Resistive thickness gauge, principle.

(1-1)3, the current flow pattern will be nonuniform, and the resistance between the contacts will be larger than that expressed by Eq. (1). The additional resistance ("spreading resistance") can be computed in simple cases.[1]

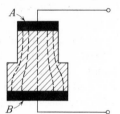

FIG. (1-1)2.    Nonuniform current flow pattern caused by abrupt change of the cross section of the test object.

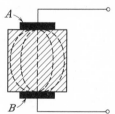

FIG. (1-1)3.    Nonuniform current flow pattern caused by electrodes that are smaller than the test object.

Surface conduction effects can be eliminated by means of the guard-ring arrangement illustrated in Fig. (1-1)4. This consists of a circular electrode $A$ (guarded electrode: radius $r_1$) on one side of the sample, concentrically surrounded by an annular electrode $B$ (guard electrode: inner radius $r_2$, outer $r_3$). The gap $r_2 - r_1$ should be small. On the other side of the sample is a circular electrode $C$ (unguarded electrode: radius $r_3$). The guard-ring electrode $B$ prevents the flow

[1] W. Schrader, *Wied. Ann.*, **44**, 222 (1891), and W. B. Kouwenhoven and W. T. Sackett, Jr., *Welding Research Suppl.*, **14**, 466s (1949).

of the surface current $i_s$ through the meter $M$. The meter (galvanometer: resistance $R_m$) measures, therefore, only the current passing through the test object. A resistor $R_p$ is provided to protect the meter in case of a flashover. If $R$ is the resistance between the electrodes $A$ and $C$ as indicated by the meter $[R = E/I - (R_p + R_M)]$, the thickness of the test object can be found from

$$t = \frac{R(r_1 + r_2)^2 \pi}{4\rho}$$

The electrodes should be in intimate contact with the material to be investigated. Mercury makes good contact but is difficult to handle. Electrodes of platinum, gold, or silver, which are deposited chemically by cathode sputtering or by means of a metal sprayer, give good

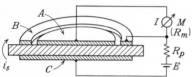

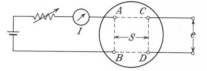

FIG. (1-1)4. Guard-ring arrangement for the separation of surface and volume resistance.

FIG. (1-1)5. Contact square for thickness measurement of a plate when only one side is accessible.

results. The use of colloidal graphite ("Dag") has been suggested. For many purposes fine tin or aluminum foil held in place by a thin layer of paraffin oil or petroleum jelly is satisfactory.

*b. Contact on One Side.* The following method permits the determination of the thickness of a metal plate if only one side is accessible.[1] The method is based upon the difference of the current flow pattern in a thin sheet (two-dimensional flow) and in a thick sheet (three-dimensional flow).

Current is passed through the plate under investigation between two contacts, $A$ and $B$, Fig. (1-1)5, and the potential difference between two other contacts, $C$ and $D$, is measured. In a thin sheet the potential difference for constant current depends upon the geometry but not upon the actual spacing of the electrodes; in the thick sheet, the potential difference for constant current is inversely proportional to the linear dimensions. In a plate of finite thickness $t$ the conditions are approximately those of a two-dimensional flow if the electrode distance $s$ is large, i.e., for $s/t \gg 1$; in this case the potential difference for similar electrode arrangements is independent of the electrode spacing. However, if the electrode distance is small, the

[1] A. G. Warren, *J. Inst. Elec. Engrs.* (*London*), **84,** 505 (1939).

flow pattern approximates that of three-dimensional flow in a thick sheet, i.e., the potential difference rises rapidly when $s/t$ becomes less than unity.

A set of "contact squares" is used. Each contact square contains in an insulator four resiliently mounted contacts (phonograph needles), as shown in Fig. (1-1)5. A set consists of squares with dimensions decreasing by a factor of 2; i.e., $s = 4, 2, 1, \frac{1}{2}, \frac{1}{4}, \frac{1}{8}$ in. First, a large contact square is pressed against the plate to be examined; the current is adjusted to a convenient value and maintained constant thereafter. The potential difference is noted. Then the next smaller contact square is used. If the size of the large contact square is considerably larger than the plate thickness, the potential difference for the same current will be only slightly higher with the smaller contact square. However, if the size of the larger contact square approaches the plate thickness, then the ratio of the potential difference $e_1$ obtained with a (smaller) square of the size $s$ to $e_2$, the potential difference obtained with a (larger) square of the size $2s$ increases rapidly. The ratio of these potential differences $e_1/e_2$ is noted, and the thickness of the plate is determined from

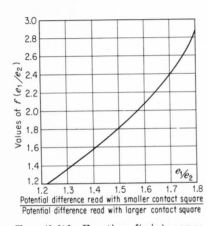

FIG. (1-1)6. Function $f(e_1/e_2)$ versus ratio of measured potential differences [*from A. G. Warren, J. Inst. Elec. Engrs. (London),* **84,** 91 (1939); *by permission of The Institution of Electrical Engineers, London*].

$$t = f\left(\frac{e_1}{e_2}\right)s$$

where $f(e_1/e_2)$ is to be found from Fig. (1-1)6, and $s$ is the side of the smaller of two successive contact squares used.

The error of the method is less than 3 per cent. The presence of paint or rust does not influence the result. The method can be used to measure the thickness of restricted plates and to detect corrosion pits. The method does not require any knowledge of the electrical conductivity of the plate material.[1]

A similar method has been described by B. M. Thornton and W. M. Thornton, *Proc. Inst. Mech. Engrs.* (*London*), **140,** 349 (1938), and B. M. Thornton, *Engineering,* **159,** 81 (1945).

[1] For underlying theory, see Warren, *ibid.*

### 1-12. Inductive Thickness Gauges

All inductive thickness gauges are based on the direct or indirect measurement of a magnetic flux. The thickness variation of the test object causes a measurable variation of the flux density. Four different cases can be distinguished:

1. The test object is of a ferromagnetic material, Fig. (1-1)7$a$. An increase of the thickness $t$ of the test object causes an increase of the

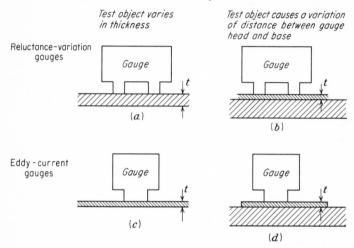

Fig. (1-1)7. Synopsis of inductive thickness gauges.

flux density. Magnetic flux meters for this application are described under Reluctance-variation Gauges (1-12$a$).

2. The test object is of a material with a magnetic permeability in the vicinity of 1 (nonferrous metal or insulator), but is, or can be, deposited on a ferromagnetic base, Fig. (1-1)7$b$. Also in this case reluctance-variation gauges of the type described under 1-12$a$ can be used, provided the ferromagnetic base is so thick that the magnetic reluctance in the circuit is primarily determined by the distance $t$ between the gauge head and the base plate, i.e., by the thickness of the test object.

3. The test object is of a conductive material having a magnetic permeability of 1. In Fig. (1-1)7$c$ it is nonferrous metal. This problem calls for the use of an eddy-current gauge, as described under 1-12$b$.

4. The test object is nonconductive (e.g., paint) but is, or can be, deposited upon a conductive base, Fig. (1-1)7$d$. This problem can also be solved with the eddy-current gauge as described under 1-12$b$.

*a. Reluctance-variation Gauges.* Examples of thickness gauges based upon reluctance variation are illustrated schematically in Fig. (1-1)8. A core $C$ of soft iron, usually laminated, carrying a coil is brought in contact with the test object $T$. The reluctance of the core is small compared to that of the test object so that the reluctance of the magnetic circuit depends primarily upon that of the test sheet, i.e., upon its thickness. Change of the thickness leads to a change of the coil inductance, Fig. (1-1)8$a$,$b$, or $c$, or of the mutual inductance between two coils, Fig. (1-1)8$d$.

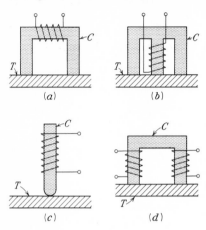

FIG. (1-1)8. Different types of reluctance-variation gauges.

Empirical calibration for each type of test material is usually required, since the reluctance depends upon the magnetic permeability of the test material. For this reason, the magnetic systems are generally used for comparison only, i.e., to detect the deviation from a standard thickness, rather than to make absolute thickness measurements.

The transfer characteristic of these gauges is usually nonlinear. The systems have been used for steel sheets of a thickness ranging from less than 0.001 to about 0.1 in., sometimes higher. The accuracy is rarely better than 2 per cent; in most cases it is about 10 to 15 per cent. The accuracy increases for thinner samples.

For references see C. E. Richards, *J. Electrodepositors' Tech. Soc.*, **14,** 101 (1938); W. H. Tait, *ibid.*, **14,** 108 (1938); W. E. Hoare and B. Chalmers, *ibid.*, **14,** 113 (1938). These references are cited by R. S. Bennett, *J. Sci. Instr.*, **26,** 209 (1949).

Several modifications of the reluctance variation-gauge system have been described as follows:

MAGNETIC-SHUNT SYSTEM. This system consists of a permanent magnet or electromagnet $A$, as shown in Fig. (1-1)9; it is brought in contact with the ferromagnetic base $B$. A part of the flux

FIG. (1-1)9. Magnetic gauge; a part of the magnetic flux is shunted by the test object $B$.

between the pole pieces passes through $B$, thus diminishing the flux in the gap $G$. The magnetic field strength in the gap $G$ can be measured with any one of the magnetic transducers described in 3-1.

The range of the system is about 0.001 to 0.750 in., the accuracy about $\pm 5$ per cent or $\pm 0.1$ mil.[1]

PERMEABILITY-VARIATION SYSTEM. The system described in Fig. (1-1)10 consists of a permanent magnet $A$ brought in contact with the ferromagnetic test object through laminated pole pieces $B$. The flux density in the magnetic circuit increases with the thickness of the test object. The magnetic permeability of ferromagnetic materials changes, in general, with the flux density. The permeability is measured by a small a-c field, superimposed on the field caused by the permanent magnet, and produced by a current in the primary coil $P$. The a-c field induces a voltage in the secondary coil $S$; the voltage is proportional to the permeability of the test object and, hence, is a function of its thickness.

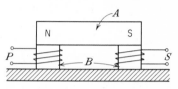

Fig. (1-1)10. Magnetic thickness gauge of the permeability-variation type.

The method has been used for the measurement of thickness from 0.01 to 0.25 in.; the accuracy varies for different constructions from 0.0005 to 0.003 in.[2]

*b. Eddy-current Systems.* A test probe consists usually of a coil wound on an insulating carrier, as shown in Fig. (1-1)11. The coil is supplied with alternating current, and the test probe is placed in contact with the metallic specimen to be examined. The magnetic a-c field causes eddy currents in the specimen which, in turn, cause a magnetic field of a direction opposite to that produced by the coil. The result is a reduction of the inductance of the coil, which can be measured. Alternatively, the test probe can contain two coils, as shown in the eddy-current displacement transducer (1-22).[3]

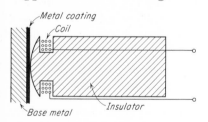

Fig. (1-1)11. Eddy-current thickness gauge.

This method requires access to one side of the sample only. The system is suitable for thickness measurements of nonmagnetic and magnetic metals as well as for nonconducting samples that are

[1] G. Keinath, The Measurement of Thickness, *Natl. Bur. Standards Circ.* 585, sec. 4.2, 1958.

[2] *Ibid.*, sec. 4.3.

[3] Application as thickness gauge by W. W. Wood, AIEE Meeting, Los Angeles, Calif., June, 1954.

backed by a metal layer.  The system can also be used for the thickness determination of a metal coating on a metal substrate, providing the conductivities of the coating and of the metal base differ sufficiently (e.g., silver on brass, nickel on steel).[1]  The depth of the induced eddy current will depend upon the conductivity of the layer through which the current flows.  The effective conductivity of the composite surface layer, and hence the magnitude of the induced eddy current, will depend upon the thickness of the coating.  Separate (empirical) calibrations are required for each combination of base metal and coating metal.

The transfer characteristic is nonlinear.  The method is capable of measuring metal coatings as thin as 2 to 3 $\mu$.  Hadley[2] has determined the thickness of aluminum films between 50 and 200 m$\mu$ by measuring the variation of the storage factor $Q$ of a coil brought close to the film.

See also M. L. Greenough, *Electronics*, **20**, 172 (November, 1947); M. L. Greenough and W. E. Williams, *J. Research Natl. Bur. Standards*, **46**, 5 (1951).  For other references see A. L. Alexander, P. King, and J. A. Dinger, *Ind. Eng. Chemistry*, **17**, 389 (1945); S. Lipson, *Am. Soc. Testing Materials Bull.*, **135**, 20 (1945); C. M. Hathaway and E. S. Lee, *Mech. Eng.*, **59**, 653 (1937).

### 1-13. CAPACITIVE SYSTEMS

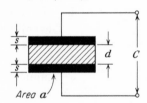

FIG.  (1-1)12.  Capacitive gauge for thickness measurement on insulators.

The capacitance between two electrodes, Fig. (1-1)12, which belong to a cross-sectional area $a$ and which are separated by a small distance $d$ (all linear dimensions in centimeters) is

$$C_{\mu\mu\mathrm{F}} = 0.0885\epsilon\,\frac{a}{d} \tag{1}$$

where $\epsilon$ is the dielectric constant of the medium between the plates. Frequently plates of circular shape (radius $r$) are used, so that for $d \ll r$, i.e., neglecting fringe effect,

$$C = 0.278\epsilon\,\frac{r^2}{d} \tag{2}$$

If the fringe capacitance cannot be neglected, the capacitance can be found from

$$C = 0.0885\epsilon\left\{\frac{\pi r^2}{d} + r\left[\ln\frac{16\pi r}{d} + 1 + f\!\left(\frac{s}{d}\right)\right]\right\} \tag{3}$$

[1] A. Brenner and J. Garcia-Rivera, *Plating*, **40**, 1238 (1953).
[2] C. P. Hadley, *Rev. Sci. Instr.*, **27**, 176 (1956).

where $s$ is the thickness of the plates. Values of the function $f(s/d)$ are tabulated in Table 1.

TABLE 1.  FRINGE-EFFECT CORRECTION FACTORS
FOR CAPACITIVE TRANSDUCERS

| $s/d$ | $f(s/d)$ | $s/d$ | $f(s/d)$ |
|-------|----------|-------|----------|
| 0.02  | 0.098    | 0.4   | 0.84     |
| 0.04  | 0.168    | 0.6   | 1.06     |
| 0.06  | 0.230    | 0.8   | 1.24     |
| 0.08  | 0.285    | 1.0   | 1.39     |
| 0.1   | 0.335    | 1.2   | 1.52     |
| 0.2   | 0.54     | 1.4   | 1.63     |

Equation (3) is strictly valid only if the capacitor plates are completely surrounded by the medium with the dielectric constant $\epsilon$.[1]

The method is applicable for the thickness determination of thin insulator layers; the minimum thickness is determined by voltage-breakdown considerations. The dielectric constant of the insulator must be known and constant. Humidity variation is likely to change the value of $\epsilon$. Sharbough and Fuoss[2] have applied the capacitive method for the determination of the width of small gaps used in breakdown studies. The (relatively large) distance $d_2$ between two semispherical electrodes is measured with a mechanical gauge and the capacitance between the electrodes is observed. The spheres are then moved closer together to the distance $d_1$, which is too small to be measured mechanically. The distance $d_1$ can be determined from the change of capacitance

$$\Delta C = 2ka\epsilon \ln \frac{d_2}{d_1}$$

where $k = 0.0885$ (for linear dimensions in centimeters), $a$ the radius of the spheres, and $\epsilon$ the dielectric constant of the medium surrounding the spheres. Distances as small as 1 $\mu$ can be determined with an error of less than 5 per cent.

Capacitive displacement transducers, which can also be used for thickness determination, are treated in detail in 1-23.

---

[1] For other capacitors see A. H. Scott and H. L. Curtis, *J. Research Natl. Bur. Standards*, **22**, 747 (1939).

[2] A. H. Sharbough and R. M. Fuoss, *Rev. Sci. Instr.*, **26**, 657 (1955).

### 1-14. THICKNESS DETERMINATION FROM MEASUREMENTS OF ELECTRIC BREAKDOWN VOLTAGE

The thickness of thin homogeneous insulating layers, such as oxide layers electrochemically produced on aluminum or its alloys, can be

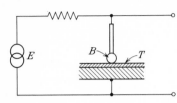

FIG. (1-1)13. Thickness gauge based on breakdown-voltage measurement.

determined from measurements of the electric breakdown voltage. The system employed for this purpose is illustrated schematically in Fig. (1-1) 13 and consists of a polished ball $B$ of copper, brass, bronze, or chromium-plated steel, about $\frac{1}{8}$ in. diameter, pressed against the test object $T$ with a force of 1 to 2 kg. An a-c voltage is applied through a safety resistor and is increased until breakdown occurs; the rms value of the voltage is measured.

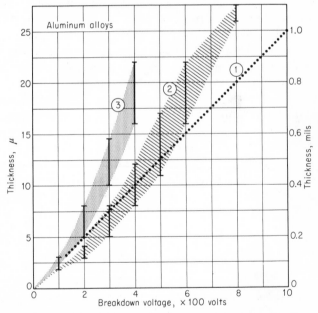

FIG. (1-1)14. Thickness of aluminum oxide coatings as a function of breakdown voltage for three different alloys (*from G. Keinath, The Measurement of Thickness, Natl. Bur. Standards Circ. 585, p. 31, 1958; by permission*).

The system has been used for the measurement of layers 1 to 25 $\mu$ thick; the breakdown voltage increases with the thickness and is about 1,000 volts for a 25-$\mu$ layer of aluminum oxide. The error is about 10 per cent. The breakdown voltage varies with the composition of the alloy. A comparative graph for different test materials is shown in Fig. (1-1)14.

For references see Anon., *Bell Lab. Record*, **20**, 278 (1942); *Am. Soc. Testing Materials, Standard B*, p. 110; J. Hérenguel and R. Segond, *Métaux-corrosion*, **20**, 1 (1945).

#### 1-15. Ultrasonic Method

The method uses an ultrasonic output transducer $T$, as shown in Fig. (1-1)15. The oscillating end of the transducer is brought in acoustical contact with the material, and the frequency of the oscillation is varied. At a certain frequency $f$ and at harmonics of this frequency, standing waves are set up in the wall; the internal damping of the material causes a sharp increase of the loading of the oscillator, and the power absorption from the transducer rises in the form of a resonance curve. The thickness $t$ of the test object can be found from

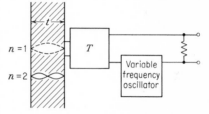

FIG. (1-1)15. Schematic diagram of an ultrasonic thickness gauge.

$$t = n \frac{c}{2f}$$

where $n$ is an integer $(1, 2, 3, \ldots)$, $c$ the sound velocity in the material of the test object, and $f$ the resonance frequency.

Steel walls ranging in thickness from 0.125 to 12 in. have been measured with this method. The accuracy is better than 5 per cent. The system permits measurement of a wall thickness from one side without requiring access to the other side, but the method requires the knowledge of $c$, the propagation velocity of the longitudinal sound waves in the material.

L. Bergmann, "Der Ultraschall," 5th ed., pp. 536ff, S. Hirzel Verlag, Zürich, 1949; G. N. Branson, *Electronics*, **21**, 88 (January, 1948); also, B. Carlin, *Electronics*, **21**, 76 (November, 1948). A summarizing paper describing different modifications of the ultrasonic method, such as pulse-echo methods, has been published by A. Lutsch and W. Böhme, *Arch. tech. Messen*, V 1124–6, May, 1957.

#### 1-16. Radiation Thickness Gauges
##### (Alpha-, Beta-, Gamma-, X-ray Gauges)

The gauge consists of a shielded source for penetrating radiation and a radiation detector such as an ionization chamber, a Geiger counter, or a scintillation counter. Variation of the thickness of the test object causes variation of the amount of radiation reaching the detector.

Two geometric configurations of source, sample, and detector have been used: that of the *absorption gauge*, Fig. (1-1)16, which requires access to both sides of the sheet to be measured, and that of the *backscatter gauge*, Fig. (1-1)17, which requires access to one side only. In the absorption gauge, source and detector are on opposite sides of the measured sheet of material, and the amount of radiation entering the detector varies with the attenuation in the sheet and hence

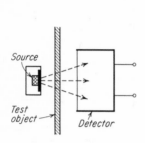

FIG. (1-1)16. Radiation thickness gauge, principle.

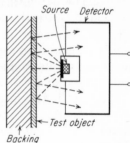

FIG. (1-1)17. Backscatter radiation thickness gauge, principle.

with its thickness. In the backscatter gauge, source and detector are on the same side of the measured sheet. The sheet must be in contact with a backing material of different atomic number; the amount of backscattered radiation from the measured sheet and the backing material depends upon the amount of scattering and absorption of radiation in the measured sheet and the backing material, and varies with the thickness of the sheet.

If a beta-, gamma-, or X-ray beam of homogeneous energy (wavelength) and of the initial intensity $I_0$ penetrates an absorbing medium of the thickness $d$, it will emerge with the intensity

$$I = I_0 e^{-\mu d} \tag{1}$$

where $\mu$ is the linear absorption coefficient.[1] For any substance, $\mu$ is proportional to the density $\rho$ of the absorbing material, i.e., the mass $m$ per unit volume $V$. The exponent in Eq. (1) is, therefore,

$$\mu d = \text{const } \rho d = \text{const } \frac{md}{V} = \text{const } \frac{m}{a}$$

---

[1] Instead of absorption coefficient, sometimes the words "extinction coefficient" or "attenuation coefficient" are used. These latter expressions are more correct, since attenuation of X-ray radiation comes about by absorption, scattering, and pair formation.

i.e., the absorption of a test specimen depends upon the mass per unit area $a$, measured in grams per square centimeter.

The linear absorption coefficient $\mu$ of a substance changes with the density, therefore, with temperature, pressure, or the state of aggregation of the absorber. The quantity $\mu/\rho$, the mass absorption coefficient (dimensions, square centimeters per gram), is independent of these variations and is customarily tabulated, therefore.

The mass absorption coefficient is different for each type of radiation (beta, gamma rays) and changes with the energy of the radiation. Gamma rays are highly penetrating; the gamma- or X-ray gauge is, therefore, applicable for heavy metals and thick specimens. Beta particles are much less penetrating and are used, therefore, for thickness measurements on thin metal sheets or foils, on light metals, and on paper, rubber, plastics, and similar substances. The alpha gauge has been used only on foils in the range of several microns. Most sources emit radiation of different kinds or of different energy levels rather than monochromatic radiation. The laws of attenuation are then more complicated than that described by Eq. (1) above, and empirical calibration is usually required.

The less penetrating the radiation is, the greater will be the change in transmitted radiation intensity for a given increment of thickness of the test object, i.e., the greater will be the sensitivity. However, the less penetrating the radiation, the lower will be the level of the intensity of radiation entering the detector and, therefore, the signal-to-noise level, i.e., the accuracy of the system.

GAMMA- AND X-RAY GAUGES. Sources of radiation are usually radioactive isotopes mounted in shielded containers. The formerly employed X-ray machines are rarely used nowadays. Several commercially available gamma sources, their gamma-ray energy levels, and their half-life times are given in Table 2. The mass absorption

TABLE 2. ISOTOPES USED FOR GAMMA GAUGES

| Isotope | Half-life, years | Gamma energy, MeV | Beta energy | Normal gauging range, mg/cm$^2$ |
|---|---|---|---|---|
| $Co^{60}$ | 5.3 | 1.33 | 0.31 | |
| $Sr^{90}$ | 20.0 | Brems radiation | 0.61 | 300–10,000 |
| $Cs^{134}$ | 2.3 | 0.56 | 0.65 | |
| $Cs^{137}$ | 33.0 | 0.66 | 0.52 | |
| $Eu^{152}$ | 5.3 | 0.12 | 0.9, 1.7 | |
| $Ra^{226}$ | 1,620.0 | 0.188 | ($\alpha$) | max 2,500 |

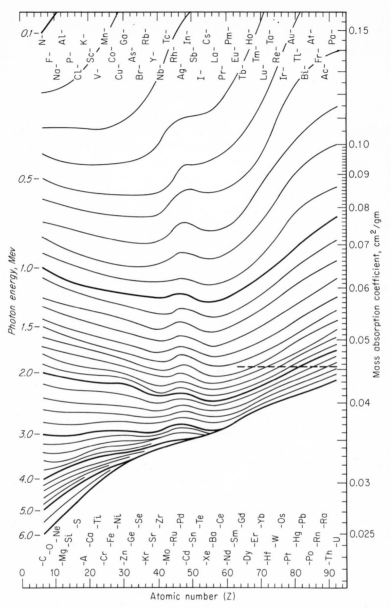

FIG. (1-1)18. Mass absorption coefficient of X rays [*from D. G. Chappell, Nucleonics*, **14**, 40 (*January*, 1956); *by permission*].

coefficient for different energy levels of the incident radiation and for different absorbers is shown in Fig. (1-1)18.

For a short summary and references on X-ray intensity and dose attenuation for inhomogeneous X-ray radiation, see F. Kohlrausch, "Praktische Physik," H. Ebert and E. Justi (eds.), vol. 2, no. 7.5112, Teubner Verlagsgesellschaft m.b.H., Stuttgart, 1956. Also see M. D. Kamen, "Radioactive Tracers in Biology," 2d ed., chap. 2, p. 61, Academic Press, Inc., New York, 1951.

BETA GAUGES. Isotopes commonly used in beta gauges together with indications of their normal gauging ranges and of the maximum

TABLE 3. ISOTOPES USED FOR BETA GAUGES

| Isotope | Half-life, years | Principal beta energy, MeV | Gamma energy, MeV | Normal gauging range, mg/cm² | Max full-scale sensitivity, mg/cm² |
|---|---|---|---|---|---|
| $Kr^{85}$ | 10 | 0.695 | 0.54 | 1–70 | 0–1.5 |
| $Sr^{90}$ | 20 | 0.61, 0.54, 0.53 | | 10–600 | 0–10 |
| $Ru^{106}$ | 1 | 0.039, 0.038 | | 200–1,200 | 0–10 |
| $Cs^{137}$ | 33 | 0.51 | 0.66 | 1–120 | 0–3 |
| $Tl^{204}$ | 3.5 | 0.78 | | 1–120 | |
| $C^{14}$ | 5,600 | 0.155 | | | |

full-scale sensitivities obtained in commercial gauges[1] are tabulated in Table 3. The absorption of beta rays approximates an exponential function over most of its range as expressed in Eq. (1). A representative absorption characteristic is shown in Fig. (1-1)19. The character of the curve can change to some extent, depending upon the geometry of the setup, the energy and energy distribution of the radiation, the kind of absorber, and the detector used, i.e., ionization chamber or electron counter.[2]

The mass absorption coefficient $\mu/\rho$ of any absorber for $\beta$ rays is nearly independent of the atomic weight of the absorber and rises only slightly

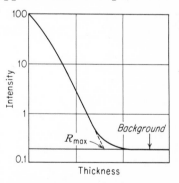

FIG. (1-1)19. Attenuation of beta rays as a function of thickness of the absorber.

[1] Values from *Information Bulletin*, Tracerlab, Inc., Waltham, Mass.

[2] For details on the attenuation of beta rays, see R. D. Evans, "The Atomic Nucleus," chap. 21, McGraw-Hill Book Company, Inc., New York, 1955.

with increasing atomic number. The mass absorption coefficient depends, of course, upon the energy of the beta radiation and can be found, for aluminum, by the empirical equation

$$\frac{\mu}{\rho} = \frac{17}{E_{max}^{1.14}}$$

where $E_{max}$ is the maximum energy in the beta-radiation spectrum, in million electron volts.[1]

Absorption-type beta gauges can be used for thickness measurements up to about 1,200 mg/cm², that is, for aluminum, a thickness of 4.5 mm. The maximum range of the backscatter gauge is about one-half that of an absorption gauge.

The radiation flux must be high enough to produce the required sensitivity, but not so high as to make shielding too unwieldy. Source strengths used in industrial applications range from 5 millicuries to 1 curie.[2]

ALPHA GAUGES. The general form of an absorption characteristic for alpha rays is shown in Fig. (1-1)20, curve $a$. The intensity of the

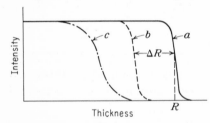

FIG. (1-1)20. Attenuation of alpha rays ($a$) in air; ($b$) after penetrating a homogeneous foil; ($c$) after penetrating a (thicker) nonhomogeneous foil.

radiation passing through an absorbing medium, such as a gas, remains unchanged until at a distance $R$ from the source (range) it decreases rapidly. The range $R$ depends upon the energy level of the radiation and the absorption in the medium; for polonium 210 in dry air at 25°C and 760 mm Hg it is about 5.5 cm.

The test object must be thin enough so that the range $R$ of the radiation, after passing the object, exceeds the thickness of the test object. The introduction of such an object in the beam decreases the range $R$ by an amount $\Delta R$ (curve $b$). The range and the range variation $\Delta R$ can be measured either by varying the distance between the sample and the detector[3] or by varying the pressure of the gas between the sample and the detector.[4] The stopping power of the object (in milligrams per

[1] M. Curie et al., *Revs. Mod. Phys.*, **3**, 427 (1931).

[2] G. B. Foster, Industrial Thickness Gages, *Nucleonics*, **14**, 66 (May, 1956).

[3] W. C. Barber, *Rev. Sci. Instr.*, **24**, 469 (1953).

[4] W. H. T. Davison, *J. Sci. Instr.*, **34**, 418 (1957).

square centimeter) is equal to that of the gas layer of the thickness $\Delta R$. Test specimens of nonuniform thickness cause an absorption characteristic as shown in curve $c$, so that thickness nonuniformity can be detected and measured.

The accuracy is about 2 per cent for 25-$\mu$ films of polystyrene and 4 per cent for 5-$\mu$ films.

With the exception of the alpha gauge, the error of radiation-gauge measurements is, in general, between 1 and 2 per cent, sometimes less. The operation is not affected by temperature or pressure variation nor by a variation of the position of the test object between the source and the detector. Only if very thin foils are measured are errors likely to arise from absorption in the air gap, which varies with temperature and pressure. The mass absorption equivalent of a 1-cm air gap is 1.18 mg/cm$^2$. Variation of the source intensity with time must be corrected.

A principal advantage of the radiation gauge is the fast and direct indication without the need for a mechanical contact between the gauge and the test object. Radiation hazards and toxicity can, in general, be avoided by adequate shielding and sealing of the source.

## 1-2. Transducers for Displacement of Force

The following transducer systems furnish an electric signal in response to a displacement or a force.

Every displacement transducer requires a force for its actuation. If the force required to produce a useful output is small or negligible compared to the forces available from the actuating system, the transducer acts as a true displacement transducer. If the displacement required by the transducer is small or negligible compared to the displacement of the actuating system, the transducer acts as a true force transducer. Practical transducers are neither pure force nor pure displacement transducers.

Force- or displacement-transducer systems are based either on resistance variation (1-21), inductance variation (1-22), or capacitance variation (1-23). A group of electrothermal systems in which the temperature and the resistance of an electrically heated wire vary in response to a displacement is described in 1-24. Systems based on the piezoelectric effect will be found in 1-25. A number of systems based upon physical mechanisms other than those described above are compiled in 1-26.

Displacements and forces can sometimes be converted into electric signals by one of the thickness-gauge systems described in 1-1. Liquid-level displacements are to be found in 1-5.

For summarizing references see H. C. Roberts, "Mechanical Measurements by Electrical Methods," The Instruments Publishing Company, Inc., Pittsburgh, 1951; and Paul M. Pflier, "Elektrische Messung mechanischer Grössen," 3d ed., Springer Verlag, Berlin, 1948. A compilation of commercial transducers can be found in D. B. Kret, "Transducers," published by Allen B. DuMont Laboratories, Inc., Clifton, N.J.

1-21. RESISTIVE DISPLACEMENT TRANSDUCERS

*a. Slide-wire Resistors.* A resistance wire or ribbon with a sliding contact, as shown in Fig. (1-2)1, represents the simplest form of a

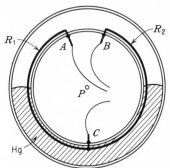

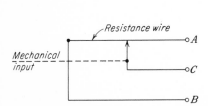

FIG. (1-2)1. Slide-wire resistance displacement transducer.

FIG. (1-2)2. Mercury-filled resistive displacement transducer (*from* W. Geyger, *Arch.tech. Messen*, V 3821-1, *November* 1935; *by permission*).

displacement transducer for the continuous conversion of longitudinal or rotational displacements into electric signals, but lends itself in general only to the construction of transducers with low resistance, up to about 10 ohms. A special mechanical construction of a potentiometer with a continuous setting permits the use of thin wires with a resistance up to 250 ohms for a 360° shaft movement.[1]

A modification of the slide-wire transducer is illustrated in Fig. (1-2)2. It consists of a doughnut-shaped glass tube containing a platinum wire (resistance about 15 ohms) or a carbon filament (100 ohms). The tube is partly filled with mercury in a hydrogen atmosphere. Rotation of the tube about its axis $P$ causes a variation of the resistances $R_1$ and $R_2$. This transducer is useful primarily for the conversion of large angles; the torque requirements are several gram-centimeters. The accuracy is limited by the surface tension of the

[1] Spiralpot, trademark, G. M. Giannini & Co., Inc., East Orange, N.J.

mercury; the error is about 1 per cent of the output for maximum rotation. The influence of temperature variation is very small, about 0.005 per cent/°C.

Higher resistance values are obtained by winding the resistance wire around an insulating core or mandrel. Figure (1-2)3 shows an arrangement for the conversion of longitudinal displacements, Fig. (1-2)4 shows the common form of the cylindrical potentiometer suitable for the conversion of limited angular displacements into electric

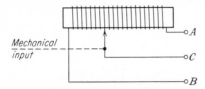

FIG. (1-2)3. Resistive transducer for longitudinal displacements.

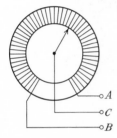

FIG. (1-2)4. Potentiometer for angular displacements; $\varphi < 360°$.

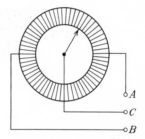

FIG. (1-2)5. Potentiometer for angular displacements.

signals, and Fig. (1-2)5 the same potentiometer for continuous rotation of the brush arm.

The resistance value of such potentiometers ranges from several ohms to several megohms, that of precision potentiometers for computers generally does not exceed the order of $10^5$ ohms.

The resistance in such potentiometers increases in steps, rather than continuously, when the contact moves from one turn of the resistance wire to the next. This stepwise variation limits the resolution of the potentiometer. The difficulty can be reduced, and the resolution can be increased, by the use of very fine resistance wire, permitting the increase of the number of turns of resistance wire per unit length of the core and thus decreasing the size of the steps. The resolution of circular potentiometers is frequently expressed by $100/n$ in per cent, where $n$ is the number of turns of resistance wire for one complete revolution (360°) of the contact-bearing shaft. The resolution of good potentiometer transducers varies from 0.5 per cent to as low as 0.05 per cent, corresponding to an angular resolution of about 2 to 0.2°. The resolution of linear motion potentiometers is usually expressed in inches or centimeters; practical potentiometers can have a resolution as low as 0.002 in. The resolution of the continuous wire potentiometer, Fig. (1-2)1, is infinitely small.

Increase of the travel length of the contact and of the resolution can be obtained by arranging the resistance and the movement of the

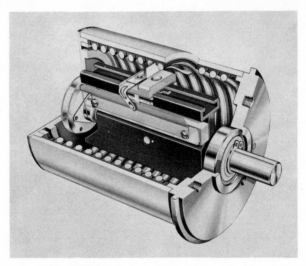

Fig. (1-2)6. Helical potentiometer (*internal view of a Fairchild Controls Corporation type* 930 *ten-turn potentiometer; by permission*).

contact brush in the form of a helix, as shown schematically in Fig. (1-2)6. Helical potentiometers with up to 40 complete revolutions of the shaft are commercially available.

The last few turns of the resistance wire on either side of the contact travel are sometimes not accessible for resistance variation. The range of angular movement through which a variation of the electrical output takes place (electrical angle) is sometimes smaller than the mechanical angle through which the potentiometer shaft can move, although potentiometers can be built with the electrical angle equal to the mechanical angle.

The resistance increases in general proportionally to the displacement or to the angle of rotation. Nonlinear functions between potentiometer output and angle of rotation can be obtained in several ways. The resistance wire can be wound on a card of varying cross sections, as shown in Fig. (1-2)7. The shape of the card should follow the derivative of the desired function. For instance, the resistance between the terminals $B$ and $C$ in Fig. (1-2)7 increases with the square of the contact displacement $d$. Another means to vary the output

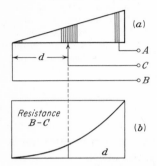

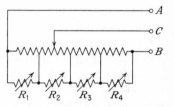

FIG. (1-2)7. (*a*) Nonlinear potentiometer; (*b*) relation between output (resistance $BC$) and displacement $d$ of the moving contact.

FIG. (1-2)8. Nonlinear potentiometer adjusted by parallel resistors $R_1$ to $R_4$.

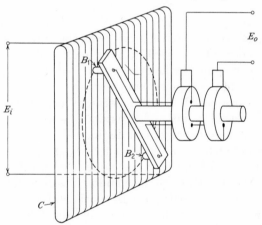

FIG. (1-2)9. Sine or cosine potentiometer.

in the form of a nonlinear function consists in an approximation of the desired function from linear segments between taps of the potentiometer, as shown in Fig. (1-2)8.

A potentiometer that furnishes an output proportional to the sine or cosine of the shaft rotation is shown in Fig. (1-2)9. The potentiometer consists of a rectangular card $C$, on which the resistance wire is wound, and two brushes $B_1$ and $B_2$ moving in a circle relative to the card. The accuracy is about $\frac{1}{2}°$.

For details concerning sine, cosine, and other nonlinear potentiometers, see F. E. Dole, in John F. Blackburn (ed.), "Components Handbook," M.I.T. Radiation Laboratory Series, vol. 17, secs. 8.5 and 8.9; F. B. Berger and E. F. MacNichol, Jr., in Britton Chance, F. C. Williams, V. W. Hughes, D. Sayre, and E. F. MacNichol, Jr. (eds.), "Waveforms," M.I.T. Radiation Laboratory Series, vol. 19, sec. 12.4; D. MacRae, Jr., and W. Roth, in Ivan A. Greenwood, Jr., J. Vance Holdam, Jr., and Duncan MacRae, Jr., "Electronic Instruments," M.I.T. Radiation Laboratory Series, vol. 21, sec. 5.7, all McGraw-Hill Book Company, Inc., New York, 1948.

The torque requirement to start the movement of the brush in precision potentiometers is generally between 10 and 1 oz-in., and it takes about half this amount to keep the brush in motion. Special potentiometers are commercially available which require not more than 0.003 oz-in. (200 dyne-cm) to move the brush. Intermittently acting torque-free slide-wire resistances can be made by means of a contact arm which moves over the resistance track, separated from the track by a short distance, i.e., without touching the resistance wire, and which is pressed against the resistance wire at periodic intervals.

A torque-free potentiometer for use in a-c circuits in which the contact arm moves at a short distance from the slide wire to which it is capacitively coupled is described by Dimeff and Fryer.[1] Over an angular range of $\pm 30°$ the error is within $\pm 0.02°$.

Nonlinearity of potentiometers is largely a matter of definition.[2] The deviation from the best-fitting straight line which can be drawn through the actual points of resistance is called "normal," or "independent," nonlinearity. It is distinguished from zero-base nonlinearity (deviation from the best-fitting straight line passing through the zero point) and from terminal nonlinearity (deviation from a straight line passing through both end points of the electric travel). The deviation from linearity in high-precision potentiometer transducers can reach values as low as 0.05 per cent, but it is generally of the order of 0.5 per cent of the total resistance value. The higher the total resistance, i.e., the thinner the resistance wire is, the higher the linearity that can be obtained. In some potentiometers nonlinearity occurs primarily in the region near the ends.

If the actuating force is given, the dynamic response of a linear

[1] J. Dimeff and T. B. Fryer, *Electronics*, **30**, 143 (February, 1957).
[2] See D. C. Duncan, Characteristics of Precision Servo Computer Potentiometers, AIEE Conference on Feedback Control Systems, Atlantic City, N.J., Dec. 6, 1951.

potentiometer is limited primarily by the mass of the moving part. In a circular potentiometer the dynamic response is limited by the moment of inertia of the rotating part, which is in general between 0.08 to 0.3 oz-in.$^2$ (15 to 60 g-cm$^2$). Special potentiometers are available with a moment of inertia as low as $2 \times 10^{-4}$ oz-in.$^2$ (0.037 g-cm$^2$). The highest frequency response is in general around 3 cps.

Noise in potentiometers arises from several causes.[1] When the brush moves over the wire, the electric contact area varies in irregular fashion. This resistance variation gives rise to a noise voltage, not only when a current is drawn from the brush, $i_o$ in Fig. (1-2)10, but also because a current $i_b$ passes from one turn of the resistance wire through the brush to the next turn. Noise can also arise from tribo-electric effect, i.e., a discontinuous electromotive force arising between two dissimilar metals in a relative motion. This noise voltage can reach values as high as 1 mV, but with the proper selection of contact metals, it can be kept below 10 $\mu$V and is usually negligible. The two types of noise described above can

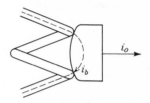

Fig. (1-2)10.   Current-flow diagram between two windings and a contact brush in a potentiometer.

sometimes be reduced, in experimental work, by the application of a very small amount of petroleum jelly or kerosene over the contact area. Noise, furthermore, stems from the discontinuity occurring when the brush in moving makes contact with a turn of the resistance wire and breaks contact with the preceding turn (resolution noise). Foreign particles getting between the contacts and oxide films or chemical alteration of the wire surface can cause considerable noise. A very serious source of noise can be contact chatter caused by a too fast movement of the brush over the contact area.

The temperature coefficient of potentiometers is in the order of 0.002 to 0.015 per cent/°C. Because of the use of different metals for the resistance wire and the brush, thermoelectric forces are likely to arise in potentiometers.[2]

The main advantage of potentiometer transducers is their accuracy and their simplicity of operation. A voltage source and a meter are in general all that is required for a simple indicating system. Potentiometers are hardly influenced by acceleration and vibration, but they

---

[1] See I. J. Hogan, IRE Convention, West Coast, Long Beach, August, 1952; special reprint from Helipot Corporation, Newport Beach, Calif.

[2] For a discussion of the technical characteristics of potentiometer transducers, see L. A. Nettleton and F. E. Dole, *Rev. Sci. Instr.*, **17**, 356 (1946).

require precision machining and, with the exception of special construction, their resolution is limited. The mechanical force and torque requirements are not always negligible. The transducers are primarily applicable for large linear movements in the order of inches or for angular movements of at least ten or more degrees. The lifetime is in general at least $10^6$ full cycles of movement of the contact shaft over the potentiometer travel. A potentiometer filled with a lubricant liquid, manufactured by the Helipot Corporation, has a lifetime of $10^7$ shaft revolutions.

 *b. Spring Transducer.* A spring is wound so that its initial tension increases uniformly over its length, Fig. (1-2)11. It may be wound as

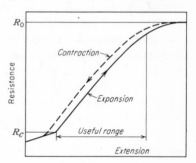

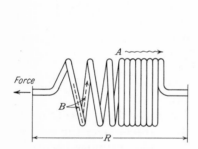

Fig. (1-2)11. Spring transducer.

Fig. (1-2)12. Calibration curve (resistance versus extension) of a spring transducer (*private communication from W. A. Wildhack*).

a cylindrical spring either with variable feed angle or variable initial tension, or from wire with uniformly increasing cross section; or it may be wound in the form of a tapered, i.e., cone-shaped helix. When the spring is closed it represents, electrically, a cylinder surface, and the current passes from one turn to the next in the direction of the arrow $A$. When a force is applied which elongates the spring, the single windings become separated one by one; the current then passes through the separate turns in the direction of the arrow $B$.

 The resistance increases with the elongation, as shown in Fig. (1-2)12; the ratio of the "open" to the "closed" resistance for gold-plated high-resistivity springs may be as high as 50 to 1.

 Spring transducers have been applied for elongations varying from 0.006 to 0.6 in. for full-scale operation. The forces required for such displacements range from less than 1 to 100 grams, and the application for higher forces presents no difficulty in principle. Displacements as small as $10^{-5}$ in. (0.25 $\mu$) may be measured with simple equipment and without the use of amplifiers.

The resistance is low. When closed, resistances of different transducers vary between 0.1 and 5 ohms; when open, the resistance values range between 3 and 20 ohms. Resistance changes between open and closed position as large as 15 ohms have been observed.

The error for short-time operation is in the order of 1 to 2 per cent; over small ranges of operation and with selected gauges better values can be obtained. The deviation from linearity of the output characteristics is $\pm 2$ per cent over limited ranges of the order of 0.001 to 0.01 in. Considerable extension of the linear range may be obtained by the use of pairs of matched gauges in a resistance bridge. The dynamic response varies widely with the spring construction. Natural frequencies of 250 cps have been observed. The (short-time) hysteresis can be considerable; it may vary from less than 1 to 10 per cent.

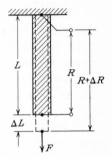

Spring transducers may be made from noble metals which are hard enough to have the elastic characteristics of a spring (platinum-rhodium, or "Paliney" alloy, primarily a gold-silver alloy) or from non-noble metals such as steel which, for better contact, should be gold-plated. The wire diameter of practical transducers is between 0.005 and 0.015 in.

FIG. (1-2)13. Schematic diagram of a resistance-wire strain gauge.

The spring transducer is, in principle, an elegant solution of the transducer problem. It is simple and sensitive, its force requirements are small compared to the wire strain gauge, the impedance is low, and it has the added advantage of requiring only a minimal amount of accessory equipment; but—like all transducer systems based on contact variations—it shows a considerable degree of unreliability, which limits its application.

*c. Wire Strain Gauges.* This strain gauge, Fig. (1-2)13, consists essentially of a fine wire, usually of about 0.001 in. diameter. When exposed to strain within the elastic limit, two physical mechanisms will cause a change of its resistance. First, its geometrical form will vary, i.e., the wire will show an increase of length and a decrease of cross-sectional area. Second, a change of electrical resistivity will occur, which can result in an increase or decrease of resistance, depending upon the material of the wire.

The strain coefficient of resistance is

$$\frac{dR}{R}\,d\epsilon = (1 + 2\mu) + \frac{d\rho}{\rho}\,d\epsilon \tag{1}$$

where $R$ is the resistance of the wire, $dR$ is the resistance variation due to the variation of strain $d\epsilon$, $\mu$ is the Poisson ratio, and $\rho$ the resistivity. Of the two terms of the right side of Eq. (1), the first denotes the geometric effect, the second the physical effect of resistivity variation. Apparently, the change of resistivity has its origin in a variation of the number of free electrons and of their mobility.[1]

In the strain-gauge engineering literature, the behavior of the strain gauge is commonly expressed by

$$\frac{\Delta R}{R} = S \frac{\Delta L}{L} \tag{2}$$

where $L$ is the length and $R$ the resistance of the unstrained wire, $\Delta L$ its change of length, and $\Delta R$ its change of resistance caused by external stress. The dimensionless magnitude $S$ (sensitivity factor of the wire) can be positive or negative and varies for different metals between $-12.1$ and $+3.6$. Table 4 shows some representative values of $S$.

The resistivity of the different wire materials changes with temperature; therefore, temperature variations will cause errors in the use of wire strain gauges. Figures for the temperature-resistivity coefficient are given in column 4 of Table 4.

TABLE 4.    STRAIN SENSITIVITY $S$ FOR WIRES OF DIFFERENT MATERIALS

| Material | Composition | $S$ | Temp. coeff. of resistivity, $1/°C$ |
|---|---|---|---|
| Manganin | Cu 84, Mn 12, Ni 4 | 0.3 to 0.47 | $0.01 \times 10^{-3}$ |
| Constantan (Advance) | Cu 60, Ni 40 / Cu 55, Ni 45 | 2.0 to 2.1 | $\pm 0.03 \times 10^{-3}$ |
| Nichrome | Ni 80, Cr 20 | 2.1 to 2.3 | $0.1 \times 10^{-3}$ |
| Nickel | Pure | $-12.1$ | $6.7 \times 10^{-3}$ |
| Iso-Elastic (cold-worked Elinvar) | Ni 36, Cr 8, Mo 0.5, balance iron | 3.6 | |
| Alloy 479 (Sigm. Cohn Corp., New York) | Pt 92, Wo 8 | $4 \pm 10\%$ | $0.24 \times 10^{-3}$ |

The effect of resistance variation with strain is small; the resistance of most wire strain gauges varies by not more than 1 per cent for a practically useful range of applied strain.

[1] G. C. Kuczynski, *Phys. Rev.*, **94**, 61 (1954); see this paper for references on recent work concerning the physical mechanism of strain gauges.

Two different forms of wire strain gauges are employed, the unbonded and the bonded strain gauge.

UNBONDED STRAIN GAUGE. An example of an unbonded strain gauge is shown in Fig. (1-2)14. Four sets of strain sensitive wires, electrically connected to form a Wheatstone bridge, are mounted under stress between a frame $F$ and a movable member $M$. If the element $M$ moves through an angle around the pivot point $P$ (dotted line), the wires $B$ and $C$ will be elongated and the wires $A$ and $D$ will

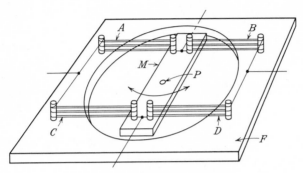

FIG. (1-2)14. Unbonded strain gauge.

be shortened. The changes of resistance will cause an unbalance of the bridge and give rise to an output signal.

The range of movement of practical unbonded resistance strain gauges is in the order of $\pm 0.0015$ in. A typical resistance strain gauge will require a force of about 0.15 oz (5 g, or $5 \times 10^3$ dynes). The resistance of unbonded strain gauges ranges from about 60 to 5,000 ohms; special strain gauges have been built with resistances in the vicinity of 50,000 ohms. The resistance variation caused by the maximum practical displacement is about 1 per cent. The magnitude of the output signal depends, of course, upon the applied voltage; for an input voltage of 10 volts, i.e., near the maximum safe limit, one can expect an output of the order of 10 to 100 mV.

Increase of the applied voltage leads to an overheating and breakdown of the wire. Forced-air cooling is not effective and causes uneven distribution of temperature. Cooling in a helium atmosphere at reduced pressure has been recommended and permits a fourfold increase in current.

The deviation from linearity of unbonded strain gauges is less than 1 per cent; the accuracy can be better than 0.1 per cent. About 1 min is usually required, after excitation has been applied, to reach thermal equilibrium. The zero point drifts, in general, by about 0.05 to

0.1 per cent of the full-scale output, apart from any long-term zero drift. The dynamic response of the unbonded strain gauge depends largely upon the mass of the moving parts to which the wires are connected. The natural resonance frequency of the wires can be in the range of several 10 kc, but practical strain gauges have usually

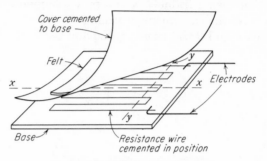

FIG. (1-2)15.  Bonded strain gauge.

a natural resonance frequency of not more than several hundred cycles per second.

BONDED STRAIN GAUGES.  The bonded strain gauge (E. E. Simmons, Jr., U.S. Pat. 2,292,549, 1942) consists of a grid of fine wire or foil cemented to a paper support about 0.003 in. thick, as shown in Figs. (1-2)15, (1-2)16, and (1-2)17. The wire is held in position by

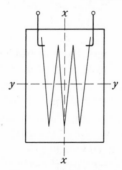

FIG. (1-2)16. Resistance-
wire strain gauge.

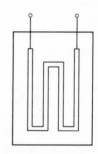

FIG. (1-2)17. Foil
strain gauge.

nitrocellulose cement or phenol resin polymerized at high temperature. Two connecting leads are either soldered or welded to the fine wire, and a piece of felt is frequently laid over the grid for thermal protection. For application, the entire strain gauge is cemented to the metal structure to be investigated, and the resistance variations are measured.

The relationship between the fractional resistance variation and the applied strain $\epsilon$ is linear.

$$\frac{\Delta R}{R} = G\epsilon$$

The gauge factor $G$ is determined from measurements under standard conditions, i.e., by cementing the bonded strain gauge to a specimen with a Poisson ratio of 0.285 and subjecting it to a strain in the specimen along the gauge axis. The deviation from linearity is in the order of 1 per cent.

The maximum range of displacement is usually of the order of $3 \times 10^{-3}$ in. ($\sim 70\ \mu$); the forces needed to produce maximum output are in the order of 10 to 100 oz.

The gauge resistance is, in general, between 50 and several thousand ohms. The practically obtained resistance variations are in the order of 1 per cent for a maximum range of displacement. The output-voltage level depends, of course, upon the input voltage used to measure the resistance variation; for an input voltage of 10 volts one can expect an output voltage between 10 and 100 mV. An accuracy as high as 0.1 per cent can be achieved.

The frequency-response characteristic extends from zero to many thousands of cycles per second, although most practical gauges have an upper frequency limit around several hundred cycles. Under sustained stress there can be a plastic deformation which corresponds to a stress up to $3.5 \times 10^{-5}$. Of this creep, 75 per cent will arise within the first 24 hr.

Since a certain length of wire is arranged in a direction $y$, Figs. (1-2)15 and (1-2)16, perpendicular to that of principal stress $x$, the wire strain gauge exhibits an effect of cross sensitivity. The cross-sensitivity factor, i.e., the ratio $k = $ transverse sensitivity/longitudinal sensitivity, is in general less than 2 per cent.[1] Gauges without transverse sensitivity have been described by Gustafsson and Huggenberger.[2] The gauges consist of parallel stress-sensitive wires imbedded in a flat strip of transparent plastic and jointed alternatively at the top and the bottom with heavier (not strain-sensitive) connectors.

The largest single source of error in using a resistance strain gauge is the effect of temperature fluctuations. Temperature variation has a twofold effect on resistance strain gauges; first, it will alter the electrical resistance of the gauge, and second, it will cause a difference

[1] R. Baumberger and F. Hines, *Proc. Soc. Exptl. Stress Anal.*, **2** (1), 113 (1944).
[2] G. V. A. Gustafsson and A. H. Huggenberger, G-H, Tepic, or Sweden Gauges, *Natl. Bur. Standards Circ.* 528, p. 79, 1954.

in expansion of the gauge wire and of the mounting, so that a strain is caused in the wire, which is superimposed to that caused by the actual displacement under measurement. Therefore, the gauges and the mechanical equipment under investigation must be kept at a constant temperature, or temperature compensation methods must be applied. The most common method of temperature compensation is to use a second strain gauge (dummy gauge); both gauges are exposed to the same temperature, but only one is exposed to the displacement under investigation. Temperature compensation is of minor importance in dynamic investigations. Thermoelectric effects arising at the lead connections may also cause errors.

Humidity entering the gauge system is likely to cause disturbances by creating insulation defects and corrosion of the wire. A protective coating is sometimes required.[1]

Magnetorestrictive effects in strain gauges consisting of ferromagnetic wires have been observed by Vigness.[2] The effect, which can cause disturbing voltages in the order of 2 mV, is not observed in new strain gauges but appears after the gauges have been "conditioned" by the application of a current and exposed to repeated strain impacts.

Gauges cemented with nitrocellulose cement can be used without appreciable error at temperatures which range as high as 60 to 80°C; gauges made from cured synthetic resin can be used up to 200°C. Good adherence of the strain gauge to the specimen under investigation is important. Nitrocellulose cement dries in air at room temperature (minimum time for drying, 3 to 4 hr; drying time for better results, 10 hr) and can be used at temperatures below 80°C. Phenol resin requires polymerization at elevated temperature (95°C for 12 hr) but is more stable and can be used up to 250°C. Air bubbles and cavities between the specimen and the strain gauge are likely to cause considerable errors. Errors may arise through hysteresis and creep. Commercial gauges are available in lengths from $\frac{1}{8}$ to 6 in. and in widths from $\frac{1}{16}$ to about $\frac{1}{2}$ in.

Strain gauges made by evaporating metals on an insulating substrate have been investigated by Campbell.[3] Satisfactory results

[1] Details on cementing and waterproofing are described by R. G. Boiten, Characteristics and Applications of Resistance Strain Gauges, *Natl. Bur. Standards Circ.* 528, 1954.

[2] I. Vigness, *Rev. Sci. Instr.*, **27**, 1012 (1956); see also J. M. Krafft, *Proc. Soc. Exptl. Stress Anal.*, **12** (2), 173 (1955), and A. Meitzler, *Rev. Sci. Instr.*, **27**, 56 (1956).

[3] W. R. Campbell, *Natl. Bur. Standards Circ.* 528, p. 131, 1954; also see A. Krimsky and R. L. Parker, *Natl. Bur. Standards Rept.* 5139, February, 1957.

were obtained with films of Pd, Pt, Sb, Co, Au, Ni, and Te deposited on a strip of anodized aluminum which was coated with silicone resin. With decreasing thickness of the film, the sensitivity (gauge factor) generally decreases first, i.e., becomes less than that of a strain gauge in bulk (wire) form, and then rises above that of the bulk material. Sensitivities 10 to 20 times higher than that of wire strain gauges have been observed. Some of the film strain gauges exhibit hysteresis, drift, and nonlinearity.

Numerous modifications of the strain gauge have been proposed to extend its application to higher temperature. Nichrome wire mounted on ceramic-coated metal foil can withstand oxidation and corrosion at temperatures up to 1000°C. The gauges can be mounted to the test specimens by high-temperature cements, such as lead silicate, silica oxide, or alumina oxide types, or by spot welding. Glass-woven strain gauges, i.e., gauges made by braiding or weaving glass-fiber-insulated wire into a ribbon, have also been used. The application is limited; most low-melting glasses become conductive at high temperature. Gauges for high temperature usually become stable only after suitable aging, i.e., repetitive heating and cooling.[1]

The advantages of wire strain gauges are in their flexibility, their small physical size, and their high stability and accuracy. Their relatively small output impedance can be advantageous for telemetering applications. Disadvantages are the considerable force required to elongate the gauge and the small output level (compared to the capacitive or inductive displacement transducer) and, for the bonded strain gauge, the fact that an appreciable time is required for bonding the strain gauge to the specimen under test. The strain gauge has been used in connection with many instrumentation systems, such as load cells, pressure-measuring systems, torque pickups, etc.[2]

For references, see W. B. Dobie and P. C. G. Isaac, "Electrical Resistance Strain Gauges," The English Universities Press, Ltd., London, and the Macmillan Company, New York, 1948; C. C. Perry and H. R. Lissner, "Strain Gage Primer," McGraw-Hill Book Company, Inc., New York, 1954; W. M. Murray and P. K. Stein, "Strain Gage Techniques," Society for Experimental Stress Analysis, Cambridge, Mass., 1958.

*d. Semiconducting Displacement Transducers.* Two basically different types of semiconducting transducers exist, one where the multiple contact area between semiconducting particles (usually carbon)

---

[1] F. G. Tatnall, Summary Report on High Temperature Strain Gauge Resistors, Contr. NONR-845(00), Office of Naval Research, Jan. 25, 1955.

[2] See A. C. Ruge, *Natl. Bur. Standards Circ.* 528, p. 93, 1954.

varies when the distance between these particles is changed, and one where the resistance of a continuous film of a $p$- or $n$-type semiconductor, such as germanium, varies with stress like that of a thin metal film strain gauge. Very little information is, as yet, available on the latter-type transducer, while a considerable amount—mostly technical information—has been published on the carbon- or carbon composition-type transducer.

One form of carbon transducer is shown in Fig. (1-2)18. It consists of two or more (up to 60) carbon disks mounted between a fixed and a movable electrode. When a force $F$ is applied, the carbon disks move together by an amount $d$, and the resistance $R$ decreases.

The transfer function, $R = f(F)$ or $R = f(d)$, is approximately hyperbolic, as shown in Fig. (1-2) 19. Several workers have tried to

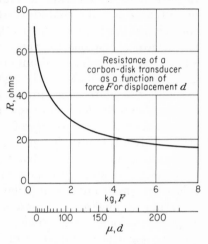

FIG. (1-2)19. Characteristic of a semiconductive (carbon disk) displacement transducer.

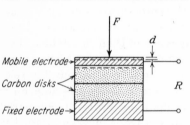

FIG. (1-2)18. Semiconductive displacement transducer.

represent this relationship by empirical equations, but none of these equations is valid over an extended range. Attempts to obtain linear transfer functions have been made; the most successful method consists of an arrangement of two transducers in adjacent arms of a resistance bridge, in which the pressure in one column is increased and that in the other decreased.

The resistivity of the carbon disks should be between $10^{-1}$ and $10^{3}$ ohm-cm; a resistivity of the order of 2 ohm-cm has been recommended.[1] The material should be hard and should have a large modulus of elasticity. The carbon disks should be thin, since the resistance variation occurs only at their surface, not within their volume.

[1] W. Glamann, *Arch. tech. Messen*, V 132–12, March, 1936.

The transducer in the form of Fig. (1-2)18 is subject to large hysteresis and drift and is severely influenced by transverse forces or displacements. The most successful method to overcome these difficulties consists of mounting the carbon electrodes, in the form of rings, on an insulated core in a rigid frame, prestressing the carbon column and using it only over a limited range of displacements.[1] The hysteresis effect which can be as much as 20 per cent of the total range is particularly high if the transducer is newly assembled but can be reduced by mechanical aging, i.e., repeated application of load.

The range of displacements for which the described form of the transducer can be used varies from about 5 to 250 $\mu$ per interface; displacements of a fraction of a micron can be detected; the range of force varies from about 100 grams to 10 kg. As in any other transducer, the force range can be considerably extended (3,000 kg)[2] through the use of mechanical load cells, and the range of displacements can be increased by the insertion of an elastic member (spring) between the moving element and the transducer. The resistance at zero load depends upon the material and the surface treatment of the carbon disks and can be between 0.5 and several hundred ohms. The variation of resistance can be as high as 10 to 1 for a variation from no-load to maximum load, but is in general much smaller.

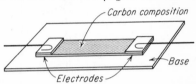

FIG. (1-2)20. Carbon-composition strain gauge.

The mechanical resonance frequency of the carbon column can be as high as 60,000 cps.[3] The mass moving with the column, such as the frame or the electrodes, reduces, of course, the useful dynamic range. Spark-plug-type pressure transducers have been built with a natural frequency between 3,500 and 11,000 cps.[4]

A carbon-composition transducer is shown in Fig. (1-2)20. It consists of an insulated carrier upon which is deposited a carbon composition such as finely divided graphite particles in a nonconducting binder (shellac, plastic, resin, cellulose acetate) or simply graphite suspended in alcohol. Two electrodes are applied. If the carrier is

[1] O. S. Peters and R. S. Johnston, *Engineering*, **116**, 253 (1923), and *Am. Soc. Testing Materials, Proc.*, **23** (II), 592 (1923); O. S. Peters and B. McCollum, *Natl. Bur. Standards Tech. Paper* 17 (247), 737 (1924); O. S. Peters, *Am. Soc. Testing Materials, Proc.*, **27** (II), 522 (1927).

[2] R. I. Martin and D. E. Caris, *SAE Journal*, **23**, 87 (1928), and *Automotive Ind.*, **62**, 230 and 592 (1930).

[3] W. Glamann and H. Triebnig, *Forschung*, **4**, 137 (1933).

[4] Martin and Caris, *loc. cit.*

subjected to stress or flexed, the resistance between the electrodes will vary. The variation of resistance occurs throughout the volume of the composition layer (not on the surface, as above).

Numerous formulas for the semiconducting mixture have been described, varying in resistance from a few ohms to the megohm range. For laboratory experiments it is sometimes sufficient to use a pencil line or a line of india ink drawn on a piece of paper.[1] Conductive rubber (rubber filled with finely divided graphite) also shows the

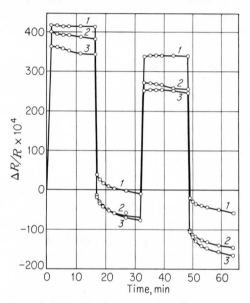

Fig. (1-2)21. Results of strain-cycle test on three semiconductive strain gauges (*from W. R. Campbell, Natl. Bur. Standards Circ.* 528, p. 138, *Characteristics and Application of Resistance Strain Gages*, 1954; *by permission*).

effect. Also ordinary carbon resistors are strain sensitive.[2] They can be used as transducers with a reproducible characteristic in the temperature range between −5 and −30°C, but are damaged if subjected to stress at higher temperature.

A study of different compositions has been made by Campbell.[3] The mixtures are painted on a piece of tissue paper which is cemented to a test specimen, similar to the arrangement of a wire strain gauge.

[1] S. L. de Bruin, *Philips Tech. Rev.*, **5**, 26 (1940).
[2] P. J. Rigden and H. J. H. Starks, *J. Sci. Instr.*, **19**, 120 (1942).
[3] W. R. Campbell, *Natl. Bur. Standards Circ.* 528, p. 131, 1954.

The sensitivity of such gauges, i.e., the fractional resistance change $\Delta R/R$ per unit strain $\Delta L/L$, varies from about 30 to 150 (highest value, for vapor carbon, 535), compared to a value of the order of 2 for wire strain gauges. The transfer function $\Delta R/R$ versus $\Delta L/L$ is, in general, nonlinear. A material which shows the least deviation from linearity consists of 9 parts, by weight, of cellulose acetate to 1 part of Dixon Micronized Graphite No. 200-08. All materials exhibit hysteresis and aftereffects, as shown in Fig. (1-2)21.

A variation of the carbon-composition transducer is shown schematically in Fig. (1-2)22. The composition is deposited to the inside or the outside of a ring-shaped elastic carrier. Four electrodes are applied so that the parts $a$, $b$, $c$, and $d$ of the conductive layers between the electrodes form the arms of a Wheatstone bridge. If a force $F$ deforms the ring, it will cause a compression in the arms $a$ and $c$ and an extension in the arms $b$ and $d$, leading to an unbalance of the bridge and causing an output signal.

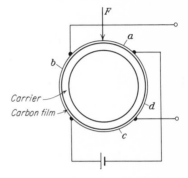

FIG. (1-2)22. Ring-type semiconductive displacement transducer (*from A. Theiss, Arch. techn. Messen,* V 1121–2 and 3, *May and June* 1943; *by permission*).

The contact resistance of the carbon disks and the resistivity of the carbon composition vary strongly with temperature. The resistance temperature coefficient is negative and of the order of $3 \times 10^{-4}$ to $10^{-3}/°C$. The above-described method of mounting two transducers in adjacent arms of a bridge reduces the temperature influence considerably. Differences of the thermal expansion between the mounting frame and the transducer also produce changes in output. The carbon-disk transducer is also sensitive to vibration and to humidity and is easily damaged by oil and vapors. Heat caused by the passage of current through the contact causes an erratic behavior.

The semiconducting transducer is an intriguing device; it has inherently many of the advantages of an ideal transducer, such as high sensitivity, small mass, good dynamic response, low actuating-force requirements, and a wide range for displacement and force. The output impedance can be varied widely to suit the requirements of the user, and the transducer itself and the associated equipment are very simple. But the semiconducting transducer is inherently unstable and shows aftereffects, hysteresis, and drift. It seems that no real improvement of the transducer can be expected from variations of the construction or of the carbon composition unless more is known about the physical mechanism causing the resistance-variation effect.

*e. Electrolytic Displacement Transducers.* The resistance between two electrodes in contact with an electrolyte depends upon the

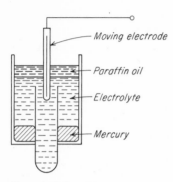

Fig. (1-2)23. Electrolytic displacement transducer [*M. Manzotti, J. Sci. Instr.*, **33**, 314 (1956); *by permission*].

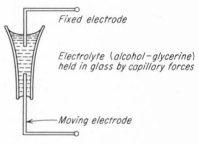

Fig. (1-2)24. Electrolytic displacement transducer [*B. S. Gunther and J. B. Concha, Proc. Soc. Exptl. Biol. Med.*, **69**, 302 (1948); *by permission*].

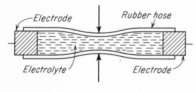

Fig. (1-2)25. Simple electrolytic displacement transducer.

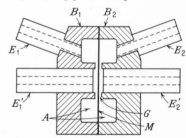

Fig. (1-2)26. Electrolytic displacement transducer used for pressure measurements. $B_1$, $B_2$, insulating body; $A$, annular chambers; $E_1$, $E_1'$, $E_2$, $E_2'$, electrodes and pressure ducts; $M$, membrane; $G$, resistance gap [*from J. R. Pappenheimer, Rev. Sci. Instr.*, **25**, 912 (1954); *by permission*].

geometry of the conducting path; any change of the length or the cross-sectional area of the conductive path causes a variation of the resistance.

A great number of electrolytic displacement transducers have been described in the literature, particularly in the field of physiology. Figures (1-2)23 to (1-2)26 show different constructions of such transducers.

Probably the most accurate electrolyte transducer has been described by Pappenheimer.[1] The system is used as a pressure-sensing device; it consists of two cylindrical pressure chambers made from quartz, with a cross section like that shown schematically in Fig. (1-2)26. The two chambers are separated by a quartz or glass membrane (0.25 mm thick, 12 to 15 mm diameter). The chambers are filled with isotonic sodium chloride solution (for biological experiments) and a trace of detergent is added to lower the surface tension. Silver tubes inserted at four places serve as electrodes and as conduit to the places where the pressure is to be measured and to a reservoir to fill the chambers. From about 70 to 98 per cent of the resistance of each cell is located in the gap $G$. The width of this gap changes when a pressure difference between both chambers occurs. The performance of this gauge is extraordinary; its lowest useful pressure range is from 0 to $\pm 2$ mm Hg, the highest range from 0 to $\pm 1,000$ mm Hg. The accuracy is high; 0.01 per cent of the full-scale range can be measured. The liquid volume of the chambers is 0.2 to 3 ml (for different embodiments), and the change of volume is in the order of $3 \times 10^{-6}$ ml/mm Hg. The output level is 0.12 to 0.4 mV/mm Hg of pressure for 1 volt applied to the bridge, the output impedance on each side, 1,000 to 3,000 ohms. The response time is short; to achieve 95 per cent of the output for a sudden change of pressure, 2 to 12 msec are required with critical damping. The influence of temperature variation is negligible if the system is used in a balanced bridge.

For the use as a displacement transducer, an electrolytic system offers a number of advantages: The transducer can be made very small, the force or energy requirements to cause a variation of the conductive path can be made negligible for many applications, the resistance of the gauge can be selected within wide limits by selecting electrolytes of proper concentration, and the resistance variations caused by a displacement of the electrodes can be large.

Electrolytic transducers suffer in general from the variation of electrolyte resistivity with temperature. The resistivity changes for most aqueous solutions by about $-2$ per cent/°C, but can change by as much as 8.9 per cent/°C (42.7 per cent sodium hydroxide solution, see temperature transducers, Table 9). A solution with an extremely small temperature coefficient is Magnanini's solution (121 grams mannite, 41 grams boric acid, 0.04 grams potassium chloride to 1 liter of solution; the temperature coefficient, at 18°C, is $-0.1$ per cent/°C; the resistivity is very high). A further difficulty, in particular if the transducer is operated at direct current, arises through polarization

[1] J. R. Pappenheimer, *Rev. Sci. Instr.*, **25**, 912 (1954).

of the electrodes, an effect which causes nonlinearity of the voltage-current characteristic. The effect can be eliminated by operation with alternating current of sufficiently high frequency, but other complications may arise through the capacitance between the electrodes and between each electrode and ground. Also, the requirements for the associated equipment (oscillator, detector, shielding) are considerable if the transducer is operated by alternating current. The described difficulties and other technical difficulties connected with most electrolytic transducers (e.g., evaporation of the solvent, capillary forces between the electrolyte and the electrodes, mechanical alterations of the rubber membranes or tubes, lack of portability) are the reasons that electrolytic transducers are primarily used for laboratory applications or where the accuracy requirements are moderate.

## 1-22. INDUCTIVE DISPLACEMENT TRANSDUCERS

An inductive displacement transducer consists essentially of a coil with the inductance $L$; a mechanical force or a displacement causes a variation of the coil parameters and consequently a change of inductance $\Delta L$. A variety of inductive transducers has been designed and described; a synopsis illustrating the more frequently used systems is shown in Fig. (1-2)27.

The inductance of a coil is

$$L = n^2 G \mu$$

where $n$ is the number of turns, $G$ a geometric form factor, and $\mu$ the effective permeability of the medium in and around the coil. Since either $n$, $G$, or $\mu$ can be changed, there exist three basic groups of inductive displacement transducers. These are described in Fig. (1-2)27, vertical columns $A$, $B$, $C$ or $D$, $E$, $F$.

*a. Inductance and Reluctance-variation Transducers.* SELF-INDUCTANCE VERSUS MUTUAL INDUCTANCE. Each transducer within a group can be designed either with a single coil, in which case the input magnitude changes the self-inductance of the coil, Fig. (1-2)27, horizontal columns 1 and 2, or with multiple coils, in which case the input magnitude causes a variation of the mutual inductance, i.e., the magnetic coupling between the coils (columns 3 and 4). A mutual-inductance system can frequently be converted into a self-inductance system by series or parallel connections of the coils. For instance, if in $B$3 the terminals $B$ and $C$ are connected, the total inductance between the terminals $A$ and $D$ will be

$$L_s = L_1 + L_2 \pm 2M$$

FIG. (1-2)27. Synopsis of inductive displacement transducers.

45

and will vary with $\varphi$ by $2M \sin \varphi$. If the two coils are connected in parallel, the inductance will be

$$L_p = \frac{L_1 L_2 - M^2}{L_1 + L_2 \pm 2M}$$

Similarly, the system $B1$ can be converted into a mutual-induction system by disconnecting contacts $B$ and $C$.

SINGLE OUTPUT VERSUS DIFFERENTIAL OUTPUT. In general the inductance variations $\Delta L$ or $\Delta M$ of a transducer will be small compared with the inductance $L$ or the mutual inductance $M$ of the system. If the succeeding stage responds to the total inductance $L + \Delta L$ rather than to the inductance variation, the sensitivity of such simple systems will be only moderate. Higher sensitivity and higher accuracy can be obtained with systems in which the input magnitude (force or displacement) simultaneously causes an increase of one inductance and a decrease of another, so that the subsequent stage measures either the difference or the ratio of two inductances (differential, push-pull, balanced, or ratio output). Each inductive transducer system within a basic group can be designed either with single output (vertical columns $A$, $B$, $C$), or with differential (ratio) output (columns $D$, $E$, $F$). The transformation from single to differential output is frequently only a matter of circuitry (see, for instance, $A1$ and $D1$ or $A2$ and $D2$).

Inductive displacement transducers with differential or ratio output offer, in general, advantages besides that mentioned above. Their output is less influenced by external magnetic fields and by temperature changes and variations of the supply voltage and frequency.

AIR- VERSUS IRON-CORE COILS. An inductive displacement transducer system can be further modified by designing it either with coils in air (horizontal columns 1 and 3) or with cores of high permeable materials. Air-coil systems can be operated at a high carrier frequency so that they are applicable to the conversion of mechanical movement over a wide frequency range. A practical advantage of iron-core systems is that they are less likely to cause an extended magnetic stray field and they are also not affected as much by external magnetic fields, as are air coils. A disadvantage of iron-core transducers is that their inductance is likely to vary with the current passing through the coil. The frequency of the supply voltage used for the measurement of the inductance should not exceed 20,000 cps for iron-core transducers. Line voltage (60 or 400 cps) is frequently used.

A number of iron-core transducers have been described.[1] Most of these transducers are variations of the basic system illustrated in Fig. (1-2)27$C$-2, which consists of a coil wound around a core of the (uniform) cross-sectional area $a$ and the effective permeability $\mu$ (i.e., the permeability at the magnetic field strength and frequency actually used). The average length of the magnetic path in the iron core is $l$, the iron core is interrupted by an air gap of the width $d$. ($d$ should be small compared with any dimension of $a$.) The inductance of the coil is

$$L = \frac{n^2}{\mathscr{R}}$$

where $\mathscr{R}$ is the reluctance of the magnetic circuit, a magnitude analogous to the resistance in an electric circuit ($\mathscr{R} = $ magnetomotive force/magnetic flux; the infrequently used magnitude permeance is the reciprocal of the reluctance). The reluctance varies with the width of the air gap and is approximately (neglecting fringe effects and leakage)

$$\mathscr{R} = \frac{l}{a\mu'} \qquad \text{where } \mu' = \frac{\mu}{1 - (d/l)\mu}$$

In most applications the reluctance is primarily that of the air gap, the contribution of the iron core is small.

Inductive displacement transducers can be used at high temperature. For such applications, the coil forms can be made from high-temperature refractive ceramics, the armature from hollow cobalt. The transducer can be enclosed in a ceramic shell sealed with refractory cement. The core can be made of magnetic alloys, such as Hypersil or Permandur, which have a curie point above 1000°F.

A transducer of the type illustrated in Fig. (1-2)27$E$-3 is described by Hermann and Stiefelmeyer.[2] The transducer which is used for the measurement of torque has no slip rings. The voltage is supplied to a pair of coils by means of a transformer with one stationary and one rotating coil; the output voltage is picked up with a similar transformer.

Compared with the capacitive displacement transducer, the inductive transducer requires in general higher mechanical forces at the input and has a lower output impedance and a higher output power level, so that sometimes the output can drive the following stage and no amplification is required.

---

[1] See H. C. Roberts, "Mechanical Measurements by Electrical Methods," chap. IV, pp. 41–97, The Instrument Publishing Company, Pittsburgh, 1951.

[2] P. Hermann and G. Stiefelmeyer, *Arch. tech. Messen*, V 171-4 and V 171-5, March and April, 1957.

*b. Variable Differential Transformer.* Figures $(1\text{-}2)27F\text{-}3$ and $(1\text{-}2)28$ show schematic diagrams of the transducer. It consists of three coils, $P$, $S_1$, and $S_2$, and a core $C$ of magnetic material. An alternating current of a frequency varying, in practice, between 60 cps and 20 kc is fed into the coil $P$; the alternating magnetic field induces

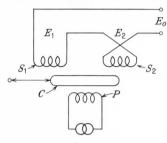

voltages $E_1$ and $E_2$ in the nearly identical coils $S_1$ and $S_2$. The output signal $E_o = E_1 - E_2$; for complete symmetry, the output signal is zero. A displacement of either one of the coils or of the core position causes a magnetic asymmetry and an output signal.

The core displacement transducer, as illustrated in Fig. $(1\text{-}2)29$, is most frequently used. The core consists usually of a nickel–iron alloy and is slotted longitudinally to reduce eddy currents.

Fig. $(1\text{-}2)28$. Linear variable differential transducer, schematic diagram.

The transfer characteristic (output voltage versus core displacement) is shown in Fig. $(1\text{-}2)30$; it is linear over a considerable range and flattens out on both ends. The output voltage changes its phase by $180°$ as the core is moved through the zero position.

The sensitivity of commercial transducers in the linear part of the transfer characteristic is of the order of 0.5 to 3 mV for a displacement of 0.001 in./volt applied to the primary coil; under standard operating conditions (6.3 volts at 60 cps) the sensitivity is between 3 and

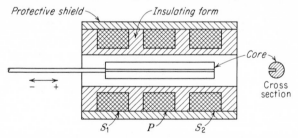

Fig. $(1\text{-}2)29$. Linear variable differential transducer, construction (*Schaevitz Engineering Corp., Camden, N.J.; by permission*).

20 mV/0.001 in. The sensitivity and the output voltage can be increased by increasing the primary voltage, but such a procedure will lead to an increase of the heating of the primary coil proportional to the square of the primary current and thus to a variation of the resistance and, in further sequence, of the output level and of the sensitivity. The effect can be reduced by feeding the primary coil

from a constant-current generator. Increase of the primary voltage also leads to an increase of the distortion of the output waveform (e.g., an increase of the applied voltage from 6 to 9 volts increases the content of higher harmonics in the output from 0.8 to 1.2 per cent; an increase to 12 volts increases the distortion to 2.25 per cent). The sensitivity also varies with the frequency of the applied a-c voltage, as shown in Fig. (1-2)31. The sensitivity varies but slowly with frequency, so that precise frequency control of the supply voltage is usually not required.

A residual voltage generally remains when the core is at the zero position, Fig. (1-2)30. This zero voltage, which is normally less than 1 per cent of the maximum output voltage in the linear range, has its origin in an incomplete magnetic or electric balance; it may consist of the fundamental frequency or

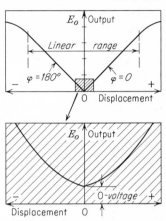

FIG. (1-2)30. Linear variable differential transformer, transfer characteristic (*Schaevitz Engineering Corp., Camden, N.J.; by permission*).

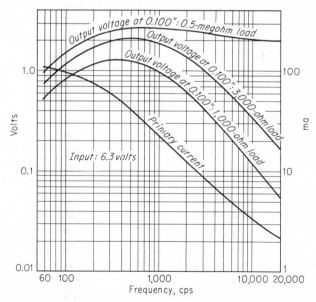

FIG. (1-2)31. Linear variable differential transformer; nominal full-range output and primary current versus frequency for the type 100SS–L (*Schaevitz Engineering Corp., Camden, N.J.; by permission*).

of higher harmonics. Frequently it can be reduced by low-pass filters in the input circuit, by balancing the individual output circuits with resistive and capacitive shunts, or by adequate grounding.

The input and output of the transducer can be interchanged. The normal connection, Fig. (1-2)28, produces higher output voltages, while the inverted connection may be preferable for matching to low impedance loads and also has the advantage of reducing the effect of stray magnetic fields (see below).

The linear-variable differential transformer may be used for a total linear range of displacements varying with the construction from $\pm0.005$ to several inches. In general, 1 per cent of the total range can still be measured; with some care it is possible to measure 0.1 per cent. The smallest displacement to be detected is of the order of $10^{-5}$ in.

The mechanical force requirements at the input depend upon the operating conditions. The axial force which the transducer exercises on the core is zero if the core is in the center position and increases linearly with the core displacement, tending to move the core toward the axial center. The force is proportional to the square of the current in the primary coil. For a commercial transducer driven at 60 cps and for full displacement, the force is of the order of 0.3 gram (300 dynes); if operated at 400 cps, it diminishes to 0.075 gram (75 dynes). The coil system also exercises a radial force on the core which increases with the square of the primary current. This force is zero when the core is in the radial center, but in this position the core is in an unstable equilibrium since the radial force increases with the radial displacement and tends to move the core toward the periphery of the bore. A restraint of this sideways movement is required.

The open-circuit output voltage for full-range displacement of the core is, in practice, between a fraction of a volt and several volts. The output impedance varies for different constructions between 20 and several thousand ohms. The output impedance is predominantly resistive, with an inductive component which varies slightly with the core displacement.

The errors of the linear-variable differential transducer amount in general to about 0.5 per cent of the maximum linear output. The deviation from linearity of commercial transducers is also about 0.5 per cent of the specified range; it can be reduced considerably by reducing the travel length of the core. The dynamic response is limited mechanically by the mass of the core (smallest mass about 0.1 gram, in larger models up to 3 grams) and is limited electrically by the frequency of the applied a-c voltage. This frequency (carrier frequency) should be at least 10 times that of the highest frequency component to be measured.

The influence of the ambient temperature upon the transducer performance is small. The effect of a temperature variation upon the magnetic properties of the core is usually negligible. Increased ambient temperature tends to increase the resistance of the primary coil and leads to a shift of the calibration curve (zero shift error) and to a reduction of sensitivity. The above-mentioned feeding of the transducer from a constant-current supply or the insertion of a thermistor with negative resistance-temperature coefficient in the primary circuit reduces the temperature effect. Commercial transducers have been built for operation in a temperature range between −50 and (intermittent) 290°C.

The operation of the transducer can be severely affected by the presence of metal masses in its vicinity and by magnetic a-c stray fields. Such fields can cause an unbalance and an excessive zero voltage as well as nonlinearity of the output. The transducer also generates a magnetic stray field which may induce magnetic disturbances in nearby magnetic masses or metal parts (by eddy currents). Such induced fields may have a back effect on the transducer. The difficulty can be reduced, but never completely remedied, by the use of magnetic shields around the transducers. Such shields should have a full-length slot to reduce eddy currents.

Linear-variable differential transformers frequently exhibit a transverse sensitivity; i.e., they respond to a radial movement of the core. The magnitude of this transverse sensitivity is in general 0.1 to 1 per cent of the axial sensitivity.

The variable differential transformer is a mechanically simple and rugged transducer suitable for displacements between 0.0001 and several inches (some authors indicate $10^{-6}$ in. as the lower limit) and forces of at least 0.1 to 0.3 gram. It has a high output level and a low or medium output impedance. The requirements for accessory equipment are moderate where an accuracy of not more than the order of 1 per cent is required, but they increase considerably for higher demands of accuracy. The dynamic response is limited by the frequency of the supply a-c voltage.[1]

A miniature inductive transducer suitable for intracardial-pressure measurements is described by Gauer and Gienapp.[2] The transducer is 12 mm long and has a maximum diameter of 3 mm. It can be used for pressures from −50 to +250 mm Hg.

[1] See *Bulletin* AA-1A, Schaevitz Engineering Corporation, Camden, N.J.
[2] O. H. Gauer and E. Gienapp, *Science*, **112**, 404 (1950).

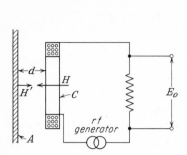

Fig. (1-2)32.  Single-coil eddy-current displacement transducer, schematic diagram.

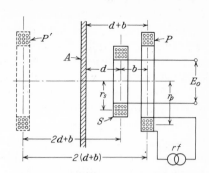

Fig. (1-2)33.  Eddy-current displacement transducer with two coils, schematic diagram.

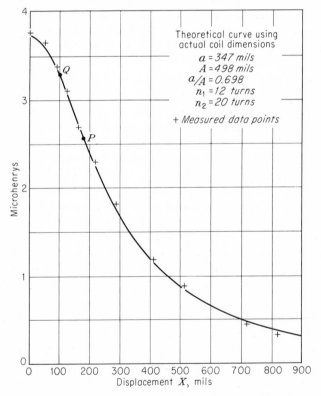

Fig. (1-2)34.  Transfer characteristic of an eddy-current transducer [*figures (1-2)34 to (1-2)36 from H. M. Joseph and P. Newman, Trans. AIEE,* **74** (I), 39 (*March,* 1955); *by permission*].

*c. Eddy-current Displacement Transducers.* The basic eddy-current transducer is shown, schematically, in Fig. (1-2)32. A current supplied by an rf generator passes through a coil $C$ and creates a magnetic field. This field produces circular and coaxial eddy currents in a conductive plate $A$ which is separated from the coil by a variable distance $d$. The eddy currents in turn cause a magnetic field which is opposed to that produced by the coil. The superimposition of the two fields reduces the effective inductance of the coil and causes a variation of the magnitude and phase of the current. The variation with separation $d$ of the current appears as a change of $E_o$ at the output terminals.

Figure (1-2)33 shows a modification of this setup. An rf magnetic field is produced by a primary coil $P$. The voltage induced in the secondary coil $S$ is

$$E_o = E_{12} - E'_{12} \tag{1}$$

or $\qquad E_o = 2\pi f I_1 (M_{12} - M'_{12}) = 2\pi f I_1 M \tag{2}$

where $E_{12}$ is the voltage induced in $S$ by the primary coil, $E'_{12}$ that induced through the effect of the eddy current, $f$ the frequency of the applied current $I$, $M_{12}$ the mutual inductance between the primary and the secondary coil, and $M'_{12}$ the mutual inductance between the image of the primary coil at the distance $d + b$ behind the surface of the plate and the secondary coil.

A typical calibration curve of the transducer is shown in Fig. (1-2)34.

The sensitivity of the transducer is

$$S = \frac{dE_o}{dx} = \omega I \frac{dM}{dx}$$

Values for the effective inductance $M$ for different geometric conditions, and for the derivative of the inductance with respect to the displacement $x$, have been computed by Joseph and Newman[1] and are reproduced in Figs. (1-2)35 and (1-2)36. They permit the computation of the sensitivity and the determination of the linear range of eddy-current transducers for different values of the radius $r_p$ of the primary coil, the radius $r_s$ of the secondary coil, and the separation $x$ between the image of the primary coil and the secondary coil. In Fig. (1-2)33, $x = 2d + b$. The sensitivity can be found from

$$S = \omega I n_1 n_2 \sqrt{\frac{r_s}{r_p}} \times [\text{ordinate of Fig. (1-2)36}]$$

$n_1$ and $n_2$ are the number of turns of the primary and the secondary

---

[1] H. M. Joseph and N. Newman, *Natl. Bur. Standards Rept.* 2558, June, 1953.

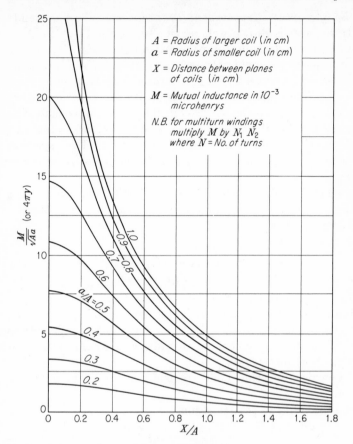

FIG. (1-2)35. Mutual inductance between two single-turn coaxial coils versus displacement between them.

coil. Practical values of sensitivity are of the order of 10 mV/0.001 in.

If the metal plate cannot be considered a perfect conductor, Eq. (1) is to be modified by the (empirical) equation

$$E_o = E_{12} - \frac{1}{1 + 0.0066\delta} E'_{12}$$

where $\delta$ is the penetration depth of the eddy current, i.e., the depth where the current is $1/e$ of that on the surface (the plate being thick enough so that the current on the back surface is essentially zero). The penetration depth is computed from

$$\delta = \frac{1}{\sqrt{\pi}} \sqrt{\frac{1}{\mu \sigma f}}$$

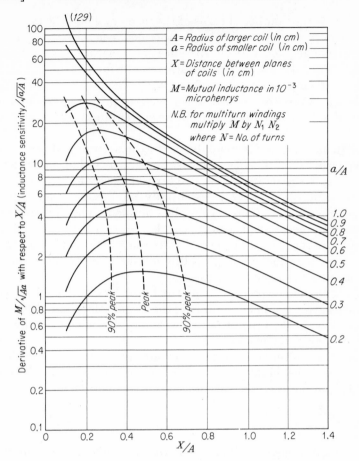

(129)

A = Radius of larger coil (in cm)
a = Radius of smaller coil (in cm)

X = Distance between planes
of coils (in cm)

M = Mutual inductance in $10^{-3}$
microhenrys

N.B. for multiturn windings
multiply M by $N_1$ $N_2$
where N = No. of turns

Derivative of $M/\sqrt{Aa}$ with respect to $X/A$ (inductance sensitivity/$\sqrt{a/A}$)

$X/A$

$a/A$

1.0
0.9
0.8
0.7
0.6
0.5
0.4
0.3
0.2

90% peak

Peak

90% peak

Fɪɢ. (1-2)36. Derivative of mutual inductance between two single-turn coaxial coils versus displacement between them.

where $\mu$ is the magnetic permeability in Mks units, $\sigma$ the conductivity of the plate material, and $f$ the current frequency.[1]

The eddy-current transducer is in general used for small displacements, i.e., for a total range from $\pm10^{-4}$ to $\pm10^{-2}$ in. Displacements of $10^{-5}$ in. can still be detected. A small repulsive force arises between the primary coil and the metal plate. For a perfectly conducting plate of an area large compared to that of the coil, the force is

$$F = -n_1^2 i^2 \frac{dM}{dx} \times 10^{-11} \text{ dyne}$$

[1] See R. W. P. King, H. R. Mimno, and A. H. Wing, "Transmission Lines, Antennas, and Wave Guides," McGraw-Hill Book Company, Inc., New York, 1945.

where $n_1$ is the number of turns of the primary coil, $i$ the instantaneous primary current, and $x$ the distance between the primary coil and the coil image (i.e., twice the distance between the coil and the metal plate). For $r_s/r_p = 1$, the value $dM/dx$ can be found from Fig. (1-2)36. For $i = 1$ amp, $n = 12$, and for $x = r_p$, the force is of the order of 10 dynes. A force can also arise between the primary and the secondary coil but is usually negligible if the current in the secondary coil is small.

Little information is available as to the accuracy of the method. Considering the difficulty of measurements at radiofrequencies, the error may be estimated to be 1 to several per cent.

The calibration curve is essentially nonlinear. Over a limited range approximate linearity can be obtained. The linearity over a selected range of displacement can be improved by increasing the radius $r_p$ of the primary coil and by decreasing the ratio of the radii of secondary to primary coil $r_s/r_p$, although such a procedure diminishes the sensitivity of the transducer.

The dynamic response is primarily limited by the frequency of the applied current, which is usually in the megacycle range. Mechanical oscillations in the range from 10 to 20,000 cps have been measured with the eddy-current transducer.

The transducer is insensitive to lateral shifts between the coil system and the metal plate, but errors are likely to occur from nonparallelism between the coil planes and the metal plate.

For references, see J. Obata, *J. Opt. Soc. Am.*, **16**, 419 (1928); H. A. Thomas, *Engineer*, **135**, 138 (1923); H. A. Thomas, *J. Sci. Instr.*, **1**, 22 (1924).

*d. Magnetoelastic or Magnetostrictive Transducers.* The magnetic permeability of ferromagnetic materials changes in general when the material is subjected to mechanical stress (Villari effect). The permeability can increase or decrease, depending upon the material, the type of stress (compression, tension, or torsion), and the magnetic flux density in the sample. Examples for the relationship between the flux density and the applied stress for iron are shown in Fig. (1-2)37; the values reported in the literature vary widely. The magnetic permeability of nickel increases when the probe is subjected to compression and decreases when subjected to tension. The general trend of the characteristics for nickel is illustrated in Fig. (1-2)38. The response to torsion of a magnetoelastic nickel probe is shown in Fig. (1-2)39, curve $a$; the flux density increases as the angle of torsion increases, regardless of the direction of rotation. A strong hysteresis effect is noticeable. If a constant longitudinal tension is superimposed to the torsion, an asymmetric curve *b* results, and beyond a critical value of longitudinal tension the response to torsion changes to that of curve *c*.

The magnetoelastic effect is particularly strong in nickel–iron alloys (63 per cent Ni, 37 per cent Fe).[1]

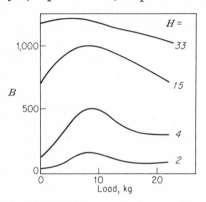

Fig. (1-2)37. Magnetoelastic effect in iron.

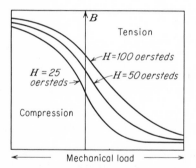

Fig. (1-2)38. Magnetoelastic effect in nickel; material subjected to compression or tension.

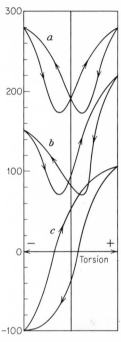

Fig. (1-2)39. Magnetoelastic effect in nickel; (*a*) nickel subjected to torsion only; (*b*) subjected to torsion plus tension; (*c*) subjected to torsion plus strong tension.

A schematic diagram and a response characteristic of a magnetoelastic transducer are shown in Fig. (1-2)40. The transducer consists of a probe *P* of magnetoelastic material around which a coil *C* is wound. If a force is applied causing a stress in the probe, the permeability of the probe and thus the inductance *L* of the coil will change as shown.

---

[1] See "Magnetostriction," The International Nickel Company, Inc., New York. For further information on magnetostriction and magnetoelasticity, see R. M. Bozorth, "Ferromagnetism," chap. 13, D. Van Nostrand Company, Inc., Princeton, N. J., 1951.

The forces required at the input of the transducer should be such as to cause in the material a stress not exceeding the order of 700 kg/cm$^2$ (10,000 lb/in.$^2$). Depending upon the cross-sectional area of the probe, forces from several grams up to several tons can be measured directly. The displacement at the input of the transducer is relatively small and will in general be in the order of several microns. The output varies between 10 and 30 per cent of the inductance at zero mechanical load, otherwise hysteresis effects may become excessive. The output inductance is in general of the order of milli-henrys up to a fraction of a henry depending upon the number of turns and the physical dimensions of the magnetic circuit.

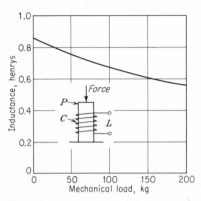

Fig. (1-2)40.  Magnetoelastic transducer, transfer characteristic; insert, schematic diagram of transducer.

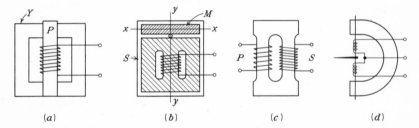

(a)          (b)          (c)          (d)

Fig. (1-2)41.  Different forms of magnetoelastic transducers: (a) magnetoelastic probe P, surrounded by shielding yoke Y; (b) magnetoelastic accelerometer [*from* H. Wilde and E. Eisele, Z. angew. Phys., **1**, 359 (1949)]; (c) mutual-inductance magnetoelastic transducer; (d) magnetoelastic phonograph pickup[*from S. R. Rich, Electronics*, **19**, 197 (*June*, 1946)].

Practical forms of transducers are shown in Fig. (1-2)41a to d. The yoke Y surrounding the magnetoelastic probe P in Fig. (1-2)41a increases the sensitivity by increasing the magnetic flux and acts as a shield against external stray magnetic fields. The yoke as well as the core should be laminated to reduce eddy currents. Figure (1-2)41b shows a magnetoelastic accelerometer as described by Wilde and Eisele. If accelerated, the mass M produces stress in the laminated core and changes its permeability; the permeability change causes the induction of a voltage in the coil wound around the central part of the core. The voltage is proportional to the variation with time of the acceleration

$$\frac{da}{dt} = \frac{d^3s}{dt^3}$$

Electric integration of this voltage furnishes a signal which is proportional to the acceleration. Magnetic shielding $S$ is used to reduce the effect of stray magnetic fields and of induction caused by the rotation of the transducer in the earth's magnetic field. The accelerometer shows a transverse sensitivity (response to acceleration in the direction $x$–$x$), an effect which can be eliminated by a mechanical construction which permits the transmission of forces from the mass to the core in longitudinal direction $y$–$y$ only. Figure (1-2)41$c$ shows a magnetostrictive transducer in which the applied force causes a change of the mutual inductance between the coils $P$ and $S$. Figure (1-2)41$d$ shows a magnetostrictive phonograph pickup containing a magnetostrictive wire $W$ (nickel, 0.02 in. diameter) arranged between the two poles of a magnet. Two pickup coils are wound around the two halves of the torsion wire. Hysteresis is reduced and linear response is obtained by giving the wire an initial torsional stress, i.e., twisting it and attaching it rigidly to the magnetic poles. A stylus displacement will increase the stress on one side of the torsion wire and decrease it on the other side by an equal and opposite amount, leading to an increase of magnetic flux in one-half the wire and decreasing it in the other half. The frequency response of this system is substantially flat over a frequency range from 150 to 15,000 cps. A practical form of a magnetoelastic pressure cell is described by Merz and Scharwächter.[1]

The response characteristic of magnetoelastic transducers is in general nonlinear. The highest frequency with which the transducer will respond is determined by the mechanical resonance frequency of the magnetoelastic probe and can be as high as several 10,000 cps. However, eddy currents are likely to introduce limitations at frequencies higher than 5,000 cps. Hysteresis errors can be large but in practical instruments can be kept at a value in the order of 1 per cent by aging, by prestressing, and by the use of the transducer over small ranges of stress variations. Some authors have observed parasitic magnetic effects which also disappear when the core is aged (subjected repeatedly to high stress).[2] Also a long-time instability

---

[1] L. Merz and H. Scharwächter, *Arch. tech. Messen*, V 132-15, November, 1937.

[2] See Wilde and Eisele, *loc. cit.*

(fatigue) amounting to as much as 2 per cent per month has been reported.

Temperature variation is of twofold influence upon the transducer performance: First, an increase of temperature causes an increase of inductance at zero stress, which is in general 0.1 per cent/°C; second, an increase of temperature changes the stress sensitivity of the transducer by about 0.2 per cent for a variation of temperature of 1°C.

Variation of the supply voltage causes relatively small variation in the transducer output, since the inductance variation depends (besides the mechanical stress) upon the magnetic field strength in the probe; it first increases with increased field strength and then decreases beyond a certain operating point. There exists, therefore, an operating region where small supply-voltage variations have no influence upon the transducer output.

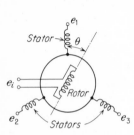

Fig. (1-2)42.   Synchro-transducer, schematic diagram.

The material for the core should have high tensile strength, and high electrical resistivity to decrease eddy currents; the maximum permeability should occur at high values of field strength to provide a high power output from the transducer.

Magnetoelastic transducers have been used primarily for the direct measurement of large forces, up to several tons, and for fast transient phenomena where frequency responses in the order of several thousand cycles per second are required. They can be built as rugged instruments that can withstand accelerations of several thousand $g$, but their characteristic depends upon temperature, and—like many transducer systems based upon variation of the physical properties of materials—they need individual calibration, i.e., their output cannot be computed and their characteristic can be changed, even irreversibly, by environmental influences.

*e. Synchros.* Synchros (trade names: Selsyn, Teletorque, Autosyn, etc.) are mutual-inductance transducers with multiple outputs and are used for rotary displacements. A synchro system with two pairs of output coils is shown schematically in Fig. (1-2)27$F$-4, another system with three output coils in Fig. (1-2)42. The synchro contains an armature or rotor with connections made through slip rings, and a stator carrying two or more (usually three) pairs of windings. In the conventional use, an a-c voltage is applied to the rotor causing a current and a magnetic flux in the rotor. The flux induces voltages in the stator windings; the magnitude of the voltages induced in each stator winding depends upon the angular position of the rotor. The voltages across each of the three stator windings are all in phase, but their amplitudes vary with the sine of the angle $\theta$ between the coil axes.

If the stator windings are arranged at 120°, the voltages are

$$e_1 = E_{max} \sin 2\pi ft \sin \theta$$

$$e_2 = E_{max} \sin 2\pi ft \sin (\theta + 120°)$$

$$e_3 = E_{max} \sin 2\pi ft \sin (\theta + 240°)$$

$E_{max}$ is the peak voltage induced in each stator coil and $f$ the frequency of the alternating voltage applied to the rotor. For a given set of stator voltages, there is only one corresponding rotor position.

Synchros can be used for the transmission of angular position information to a remote location. In a simple arrangement, the stator windings of a synchro "generator" are connected to the corresponding stator windings of a synchro "motor," as shown in Fig. (1-2)43. The generator and motor are electrically identical, but the motor is equipped with some mechanical damping. The rotors of both units are connected to the same power source; the angular displacement of the motor follows that of the generator.

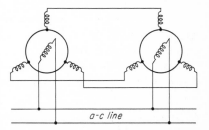

Fig. (1-2)43. Two synchro transducers acting as input and output transducers.

Synchros can be used for a range of rotation of 360°. The torque required to drive the rotor of a synchro depends upon the stator current, which varies with the external load. For a high impedance load ($>1,000$ ohms), the driving torque required is only that necessary to overcome bearing friction (order of $3 \times 10^3$ dyne-cm and higher, increasing at lower temperature) and inertial force. However, if the currents in the stator windings are of the order of $50$ mA, a torque of about $7 \times 10^4$ dyne-cm/deg is required.[1] In coupled synchro devices, Fig. (1-2)43, a torque of 3 to $7 \times 10^3$ dyne-cm is required for every degree difference in the position of the two rotors.

The output voltage (peak value) of a synchro is usually about 15 to

[1] W. F. Goodell, in John F. Blackburn (ed.), "Components Handbook," M.I.T. Radiation Laboratory Series, vol. 17, p. 328, McGraw-Hill Book Company, Inc., New York, 1949.

50 per cent lower than the input voltage and is shifted in phase against the input voltage because of leakage inductance. The sensitivity varies widely for different synchros from several hundred millivolts to the order of 1.5 volts/deg of rotation (in the zero degree position), and the accuracy varies for different synchros from $\pm 0.1$ to about 7°. The error also depends upon the rotor position and is a maximum, for the three-coil synchro, at 60°. Synchros will operate at velocities up to 1,200 rpm.

Synchros are generally rugged transducers; they are little affected by temperature in the range from $-55$ to $+95°C$ or by humidity up to 95 per cent. Special differential synchro systems can be used to form the sum or the difference of two or more angular displacements. Synchros with two perpendicular winding systems furnishing an output in the form of two orthogonal components (sine and cosine of the displacement angle) are frequently referred to as resolvers. A circuit for the use of a three-coil synchro as a resolver is described by Berger and MacNichol.[1]

A modification of the synchro system, the Magnesyn, is shown in Fig. (1-2)44. It consists of a toroidal coil wound on a core of high-permeability and low-saturation field strength, such as Permalloy. A permanent magnet in the form of a circular disk is mounted in the center of the toroid and coaxial with it. The disk can be rotated around the pivot point $P$. Two magnetic fields are induced in the Permalloy core. The permanent magnet causes a constant ("d-c") magnetic field which is sufficient to saturate the core. An a-c current of the frequency $f$ fed into the toroid windings causes a magnetic a-c field. The two fields are superimposed in parallel on one-half the toroid and in opposition on the other half. The net effect is the induction of a voltage of twice the a-c frequency alternatingly occurring in one or the other half of the toroid windings. The coil is tapped at three places; the voltage induced between two taps is

$$e = V \sin 4\pi ft \cos \theta$$

where $\theta$ is the angle between the north-south direction of the magnetic disk and the center of the coil. The operation is analogous to that of a synchro. Commercial Magnesyns require only a few watts of power and are accurate to about $\pm 0.25°$.

*f. Induction Systems.* Transducer systems of the induction type consist essentially of a coil with $n$ turns in a magnetic field, as shown

[1] F. B. Berger and E. F. MacNichol, in Britton Chance, F. C. Williams, V. W. Hughes, D. Sayre, and E. F. MacNichol (eds.), "Waveforms," M.I.T. Radiation Laboratory Series, vol. 19, sec. 12.7, p. 444, McGraw-Hill Book Company, Inc., New York, 1949.

in Fig. (1-2)45. Any change of the magnetic flux $\Phi$ passing through the coil causes, by electromagnetic induction, an output voltage $e_o = -n \, d\Phi/dt$. The output voltage is proportional to the rate of change of the magnetic field; if the magnetic field is changed in response to a physical displacement, the transducer output is proportional to the

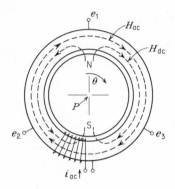

Fig. (1-2)44. Magnesyn transducer, schematic diagram.

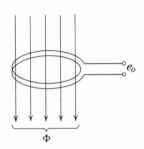

Fig. (1-2)45. Induction-type displacement transducer, principle.

time derivative of such a displacement, i.e., to the velocity. However, the output can be used directly as a measure for the displacement in those cases where the displacement and the velocity are directly related (e.g., for harmonic oscillations at constant frequency). In all other cases the voltage $e_o$ can be fed into an integrating system, and the output from the latter is proportional to the displacement. An advantage of the induction system over the systems described above is that the output voltage is generated in the transducer and no auxiliary voltage source is required. Therefore, the dynamic response of the transducer is not limited by the frequency of an

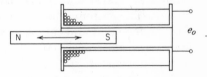

Fig. (1-2)46. Induction-type displacement transducer for longitudinal movement [*from T. A. Perls and E. Bushmann, Rev. Sci. Instr.,* **22,** 475 (1951)].

a-c supply voltage or carrier frequency, and a considerable frequency range can be encompassed. The variation of the flux can be accomplished in different ways.

One method to vary the flux developed for ballistocardiographic measurements is illustrated in Fig. (1-2)46.[1] The transducer consists

[1] See also T. A. Perls and C. W. Kissinger, *Natl. Bur. Standards Rept.* 2733, September, 1953; G. Ising, *Ann. Physik,* (5) **8,** 911 (1931).

of a permanent bar magnet, one pole extending inside of a coil with many turns. An axial relative movement between the magnet and the coil causes an output voltage.

The use of Alnico V magnet is recommended because the poles in such a magnet are short and well defined. The sensitivity is

$$S = \frac{e_o}{v} = -n'\Phi \times 10^{-8}$$

where $e_o$ is the output voltage in volts, $v$ the velocity in centimeters per second, $n'$ the number of turns per centimeter of coil length, and

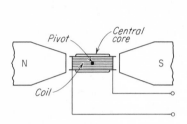

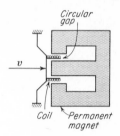

Fig. (1-2)47. Induction-type displacement transducer for rotary movement.

Fig. (1-2)48. Induction-type displacement transducer for longitudinal movement.

$\Phi$ the flux in gauss per square centimeter through a cylindrical surface surrounding one-half the magnet. Practically obtained sensitivities are about 0.01 volt/(cm/sec).

Figures (1-2)47 and (1-2)48 show two forms of moving-coil transducers. The system illustrated in Fig. (1-2)47 is similar to the usual meter movement and can be used for the conversion of rotary movements into electric signals; that illustrated in Fig. (1-2)48 is similar to the movements used in moving coil or electrodynamic speakers and can be used as a transducer for longitudinal movement. The output voltage is

$$e_o = Banv \cos \varphi \qquad (1)$$

where $B$ is the flux density in the air gap, $a$ the coil area, $n$ the number of turns, $v$ the velocity with which the coil moves, and $\varphi$ the angle between the coil plane and the magnetic-field lines. The angle $\varphi$ in Eq. (1) is usually negligible; in the system in Fig. (1-2)47 the angle $\varphi$ is zero in the position illustrated, and the angular displacement is small so that $\cos \varphi$ differs from one by an insignificant amount. In the moving-coil system with circular pole pieces and in the system shown in Fig. (1-2)48 the fields extend radially and $\varphi$ is always zero. Moving-coil systems of the type shown in these figures have been built with very small moments of inertia and can be used for a frequency range extending from a few cycles per second to more than 10 kc.[1]

The so-called variable-reluctance phonograph-pickup system is a further example of an induction-type transducer. It is illustrated in Fig. (1-2)49 and contains a permanent magnet with pole pieces $N$ and $S$ and an iron armature $A$ moving around a pivot. Two rubber blocks $B$ impart a restoring force to the armature and prevent it from sticking to the pole faces. As the armature moves, the magnetic flux through it varies and the flux variation induces in the (fixed) coil a voltage

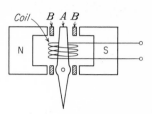

$$e_o = \frac{nMv}{d^2}$$

where $n$ is the number of turns of the coil, $M$ the magnetomotive force in the gap, $v$ the velocity of the armature tip in the gap, and

FIG. (1-2)49. Reluctance-variation displacement transducer [*from S. Kelly, J. Audio Eng. Soc.*, **2**, 163 (1954); *by permission*].

$d$ the gap width between the armature and the pole. The maximum displacement for which such pickup systems can be used is of the order of 0.1 mm; larger displacements cause a nonlinear distortion. The effective mass of the moving part can be reduced to below 5 mg, and the compliance (displacement per applied force) can be as high as $2 \times 10^{-6}$ cm/dyne. Although the armature has a resonance (in some cases around 3 to 4 kc), its effect can be practically eliminated by appropriate damping, and a fairly linear frequency response (although not flat, its output increases with frequency) can be obtained in a range from 40 to 15,000 cps. The output impedance depends on the coil construction; for high-impedance pickups, coils of about 5,000 turns are used, having a d-c resistance of 2,000 ohms and an inductance of about 1 henry. For a movement with a velocity of about 10 cm/sec at a frequency of 1 kc the output is of the order of 10 mV.

Induction systems are used frequently in displacement measurements of rotating systems in which a rotating permanent magnet induces a voltage in a stationary coil, so that no slip rings are required. A system for the measurement of torque is described by Oesterlin.[2] Two rotating magnets induce a-c voltages in two-coil

---

[1] S. Kelly, *J. Audio Eng. Soc.*, **2** (3), 169 (1954); see also E. M. Villchur, *Audio*, February, 1956, p. 40.

[2] W. Oesterlin, *Arch. tech. Messen*, V 61-3, July, 1952.

systems; the phase angle between the two voltages increases with the torsion of the axis.

### 1-23. CAPACITIVE TRANSDUCERS

The general form of a capacitive displacement transducer is shown in Fig. (1-2)50. The capacitance between the terminals (neglecting fringe effects) is $C = k\epsilon a/d$, or numerically,

$$C_{\mu\mu F} = 0.0885\epsilon \frac{a}{d} \qquad \text{linear dimensions, cm}$$

$$\text{(1)}$$

or $\qquad C_{\mu\mu F} = 0.225\epsilon \frac{a}{d} \qquad \text{linear dimensions, in.}$

where $\epsilon$ is the dielectric constant of the medium between the plates $P$, $a$ the area formed by the projection of the plates upon each other (overlapping area), and $d$ the distance between the plates.

The fringe effect (stray capacitance at the edges of the plates) cannot always be neglected. In a capacitor formed by two circular plates in which the ratio of plate radius to plate separation is 50 and in which the plate thickness is equal to the separation, the capacitance is actually 6 per cent greater than the value computed from Eq. (1). The fringe effect contributes less than 1 per cent if the ratio of radius to separation is greater than 200. (For formulas for the computation of the fringe effects see Thickness Transducers, 1-13.)

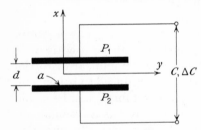

FIG. (1-2)50. Capacitive displacement transducer, schematic diagram.

The capacitance can be varied in response to a physical displacement in three ways: (1) by varying $d$, i.e., by changing the plate separation; (2) by varying $a$, i.e., moving one plate in the direction $y$, Fig. (1-2)50; and (3) by varying $\epsilon$, e.g., by moving a body with a dielectric constant higher than air into the gap between the plates. The latter method is seldom used.

*a. Change-of-distance Systems; Two-plate Capacitor.* This type of capacitive transducer is most frequently used. The response characteristic, $C = f(d)$, is hyperbolic, Eq. (1), and only approximately linear over a small range of displacement. Brookes-Smith and Colls,[1] have improved the linearity of the transducer by inserting in the gap

[1] C. H. W. Brookes-Smith and J. A. Colls, *J. Sci. Instr.*, **16**, 361 (1939).

a material with high dielectric constant (mica) which partially fills the gap. The correct thickness to obtain linearity is found by methodical experimentation.

EXAMPLE: In a capacitor with flat circular plates of $\frac{1}{2}$ in. diameter separated by a parallel gap of 0.001 in., a mica disk approximately 0.00075 in. was inserted. This caused a linear characteristic over a range of displacement of about 0.0001 in. in which the capacitance varied by 20 $\mu\mu$F.

The sensitivity of the capacitive transducer with variable separation is

$$S = \frac{\Delta C}{\Delta d} = -\frac{\epsilon a}{d^2}$$

i.e., the sensitivity is proportional to the area of the plates and inversely proportional to the square of the separation of the plates. Theoretically, the sensitivity of the transducer can be increased to any desired degree by making the distance $d$ increasingly smaller, but a practical limit is reached when the electric field strength in the air gap exceeds the voltage-breakdown limit.[1]

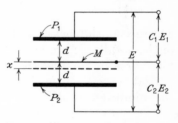

FIG. (1-2)51.  Differential-capacitor displacement transducer.

Frequently the output from a capacitive transducer is applied to a modifier system which responds to the fractional capacitance variation $\Delta C/C$ rather than to the absolute variation $\Delta C$. In this case the sensitivity is

$$S' = \frac{\Delta C}{C\,\Delta d} = -\frac{1}{d}$$

i.e., the sensitivity $S'$ increases in inverse proportion to the separation $d$ and is independent of the other dimensions of the capacitor. The response characteristic $\Delta C/C$ versus $\Delta d/d$ is linear.

DIFFERENTIAL CAPACITOR. The capacitor shown in Fig. (1-2)51 has three plates; two fixed plates $P_1$ and $P_2$ and a plate $M$ which together form the two capacitors $C_1$ and $C_2$. In the center position the distance between $M$ and either plate is $d$; the capacitors $C_1$ and $C_2$ are equal. If the middle plate is moved by an amount $x$ in response to an applied mechanical signal, the magnitude of the two output capacitors will be

$$C_1 = \frac{\epsilon a}{d+x} \qquad \text{and} \qquad C_2 = \frac{\epsilon a}{d-x}$$

[1] The breakdown strength in air at atmospheric pressure is 30 kV/cm.

Two different applications of this type of capacitive transducer can be distinguished:

1. The stage following the transducer responds to the *difference* of the partial capacitances $C_1 - C_2$. Usually an a-c voltage $E$ is applied between plates $P_1$ and $P_2$, and the difference of the partial voltages $E_1 - E_2$ is measured. The partial voltages are

$$E_1 = \frac{EC_2}{C_1 + C_2} \qquad E_2 = \frac{EC_1}{C_1 + C_2}$$

or

$$E_1 = E\frac{d + x}{2d} \qquad E_2 = E\frac{d - x}{2d}$$

Therefore, the difference of the partial voltages is

$$\Delta E = E_1 - E_2 = E\frac{x}{d}$$

The relationship between the output signal $\Delta E$ and the displacement of the middle electrode $x$ is linear and independent of the capacitor plate area or the dielectric constant. Also the magnitude of each partial voltage, $E_1$ or $E_2$, varies linearly with the displacement $x$ of the middle electrode. The sensitivity of the system is

$$S = \frac{\Delta E}{\Delta x} = \frac{1}{d}$$

If stray capacitance cannot be neglected (e.g., the capacitance in the input of the subsequent stage), the sensitivity will be reduced, and nonlinearity will arise between the displacement $x$ and the output. According to Reisch,[1] the output voltage is

$$\Delta E = E\frac{2x}{d}\frac{1}{1 + B}\left[1 + \frac{B}{1 + B}\left(\frac{2x}{a}\right)^2 + \left(\frac{B}{1 + B}\right)^2\left(\frac{2x}{a}\right)^4 + \cdots\right]$$

where $B(= C_s/C_1)$ is the ratio of the stray capacitance $C_s$ to the capacitance $C_1$ or $C_2$ if the middle plate $M$ is in the center position. For $B = \frac{1}{4}$ and $x/d = \frac{1}{5}$, the sensitivity is reduced by $\frac{1}{5}$ and the deviation from linearity is 1 per cent.

2. The subsequent stage responds to the *ratio* $C_1/C_2$ (e.g., balanced bridges, ratio meters). For such systems, the relationship between the output signal and the displacement of the middle electrode is

$$\frac{C_2}{C_1} = N = \frac{d + x}{d - x}$$

[1] S. Reisch, *Z. Hochfrequenztech.*, **38**, 104 (1931).

The output varies in a nonlinear fashion with the displacement $x$, as shown in Fig. (1-2)52. Only for very small displacements; when $x \ll d$, does

$$N = 1 + \frac{x}{d}$$

For $x/d = \frac{1}{5}$, the deviation from linearity is about 20 per cent.

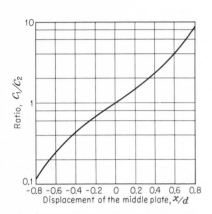

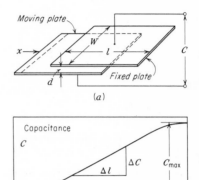

Fig. (1-2)52. Transfer characteristic $C_1/C_2$ versus displacement of the middle electrode of a differential-capacitor transducer.

Fig. (1-2)53. Capacitive displacement transducer: (a) schematic diagram; (b) transfer characteristic.

*b. Change-of-area Systems.* Figure (1-2)53a shows an example of this system and Fig. (1-2)53b a transfer function of a capacitive transducer where a longitudinal displacement of one plate causes a variation of the area $a$ and, thus, of the capacitance. In the middle part of the characteristic, the capacitance can be expressed by

$$C = \epsilon \, \frac{lw}{d}$$

where $l$ is the length and $w$ the width of the overlapping part of the two plates. The sensitivity is

$$S = \frac{\Delta C}{\Delta l} = \frac{\epsilon w}{d}$$

and the sensitivity for a fractional capacitance variation $\Delta C/C$ is

$$S' = \frac{\Delta C}{C \, \Delta l} = \frac{1}{l}$$

i.e., the sensitivity $S'$ decreases as the overlapping area increases (for a fixed width; the width has no influence upon the sensitivity). The characteristic $\Delta C/C$ versus $\Delta l/l$ is linear.

Figure (1-2)54 illustrates two capacitive transducers for rotary motion, a single and a differential transducer. The sensitivities and the transfer characteristics are analogous to those for linear movement. The characteristic $C = f(\varphi)$ can be modified by appropriate

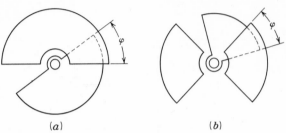

(a)                              (b)

FIG. (1-2)54. Capacitive transducers for rotary motion:
(a) two-plate capacitor; (b) differential capacitor.

shaping of one of the two plates. The sensitivity of either type (also that for longitudinal-motion transducers) can be increased by multiple-plate arrangements.

Another method to obtain high sensitivity is by the use of the serrated-type transducer shown in cross section in Fig. (1-2)55. A

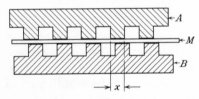

FIG. (1-2)55. Serrated-capacitor displacement transducer.

longitudinal shift of one capacitance plate by an amount $x$ against the other has the effect of changing the capacitance from the minimum to the maximum value. The transfer characteristic is similar to that in Fig. (1-2)53. A sensitivity of the order of $1\mu\mu F/0.0001$ in. has been obtained.

Capacitive transducers have been used for the measurement of extremely small displacements down to about one-tenth of an angstrom unit ($10^{-9}$ cm, $4 \times 10^{-10}$ in.), i.e., about one-tenth the diameter of a hydrogen atom.[1] On the other hand, they have been proposed for the measurement of distances between 3 and 30 m in airplane altimeters.[2] The change-of-distance method is generally preferable for the small and very large ranges; the change-of-area method is used for displacements of the order of centimeters.

[1] *Natl. Bur. Standards News Bull.*, **42** (1), 1 (1958).
[2] P. C. Sandretto, *Proc. IRE*, **32**, 167 (1944).

The force requirements can be extremely small, probably the smallest of any transducer system, with the exception of some optical systems.

The force between two capacitor plates is

$$F = \frac{8.85 \times 10^{-12}}{2} \frac{E^2 a}{d^2}$$

where $E$ is the potential difference between the plates. When a capacitive transducer has two plates of an area $a$ of 2 cm$^2$ which are 1 mm apart and connected to a potential difference of 100 volts, the force of attraction between the plates (i.e., the force required to move the plates by an infinitesimal amount) is of the order of 1 dyne.

The capacitance variations representing the output signal are generally between $10^{-3}$ and $10^3$ $\mu\mu$F; capacitance variations of $10^{-6}$ $\mu\mu$F have been observed.[1] The fractional capacitance variations are of the order of $10^{-6}$ to 1; smaller values ($2 \times 10^{-9}$) have been observed in laboratory experiments.[2]

The magnitude of the output impedance depends upon the frequency of the alternating current used for the determination of the capacitance; for practical cases (capacitances of the order of 10 to several hundred $\mu\mu$F), the output impedance is in the range between $10^3$ and $10^7$ ohms. The magnitude of both the output signal and of the output impedance may be changed by series and parallel capacitor circuits, but such modifications are always accompanied by a reduction of the signal.

The dynamic response characteristic of capacitive transducers, as well as hysteresis, mechanical aftereffects, and drift, and the influence of the environmental temperature and pressure upon the transducer performance are all determined by its mechanical construction rather than by its electrical characteristic. The source of the greatest mechanical difficulties is frequently the insulation employed to hold the plates in position.

Errors may arise from humidity affecting the insulation resistance and from stray electric fields inducing parasitic potentials in the not-grounded plate and its connections.

Cables connecting the transducer with other elements can also cause serious errors. Any capacitance variation arising in such a cable can produce a spurious signal. Further noise in such cables can arise from relative motion between a conductor and the dielectric, in

[1] R. Gunn, *Phil. Mag.*, (6) **48**, 224 (1924).
[2] W. Bürger, *Z. Physik*, **91**, 679 (1934).

particular when such a motion leads to a local separation between the conductor and the insulator. The phenomenon has been described by Perls.[1] The effect can be considerably reduced by the application of a conductive layer (carbon compound) which firmly adheres to the inside and the outside of the dielectric.

The capacitive displacement transducer is primarily useful where very small forces are available. The transducer has the principal advantage that the physical mechanism involved in its action does not depend upon any physical property of materials, with the exception of a few constructions involving the use of dielectric materials between the plates. Therefore, a high degree of stability and reproducibility can be obtained. A disadvantage of the capacitive transducer is its relatively large output impedance, which requires careful shielding and short connections to subsequent stages.

The transducer has been used for a wide variety of measurements of physical magnitudes, such as pressure, thickness, acceleration, etc., which, as far as they fall into the scope of this book, are described under the respective headings. For the application of the capacitive transducer for the measurement of temperature (on the order of $6 \times 10^{-5}$°C) see W. Sucksmith, *Phil. Mag.*, (6) **43**, 223 (1922). The construction of a microbalance with the use of a capacitive transducer is described by R. Whiddington and F. A. Long, *Phil. Mag.*, (6) **49**, 113 (1925). The smallest detectable mass (noise level) is about $10^{-9}$ gram. A capacitance-type torque meter and a capacitance transducer with toothed disks for it are described by H. G. Mills, *J. Sci. Instr.*, **25**, 151 (1948). For constructional details on capacitive transducers, see also E. A. Holmes, III, in John F. Blackburn (ed.), "Components Handbook," M.I.T. Radiation Laboratory Series, vol. 17, chap. 9, McGraw-Hill Book Company, Inc., New York, 1949.

### 1-24. THERMAL DISPLACEMENT TRANSDUCERS

*a. Bolometer Systems.* The operating principle of these transducers is the following: A spiral made of platinum wire is heated by a constant current. When the spiral is elongated, the cooling rate is increased, and thus the temperature of the spiral and, in further consequence, its resistance decreases.

A complete transducer system is shown schematically in Fig. (1-2)56.[2] Four platinum-wire spirals are electrically connected to form a bridge and mechanically suspended between posts mounted on a base plate and on a movable lever $L$; a rotary displacement of the lever in the direction of the arrow elongates the spirals $A$ and $D$ and compresses the spirals $B$ and $C$. The spirals are heated by the bridge current; their temperature varies with their elongation. The variation of temperature leads to a resistance variation and causes

---

[1] T. A. Perls, *J. Appl. Physics*, **23**, 674 (1952).
[2] See also E. B. Moss, *J. Sci. Instr.*, **7**, 393 (1930).

an output signal. No data on the performance of this transducer are reported.

  *b. Position Convectron.* A sealed glass bulb with two arms (Fig. (1-2)57, is filled with an inert gas such as argon and contains a nickel wire running centrally through each arm. The wire is electrically heated to a temperature of about 400°C; cooling occurs primarily by convection. The convection current, and thus the temperature and the resistance of the nickel wire, depend upon the angular position of

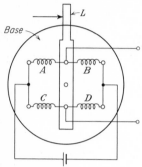

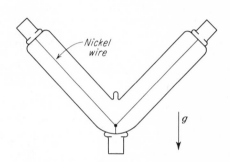

FIG. (1-2)56. Thermal displacement transducer (*after G. A. Shakespear, British Pat.* 219,-452).

FIG. (1-2)57. Thermal displacement transducer for rotary movement, convection type.

the arm with respect to the direction of gravitation (arrow); the resistance of the wire will increase in one arm and decrease in the other if the transducer is rotated from the zero position. The transducer is usually connected into a resistance bridge.

  The displacement signal can be as high as 50 mV/deg in the vicinity of the zero position. The sensitivity varies with the applied voltage and reaches a maximum at about 10 volts. Operation at direct current is recommended for the measurement of very small angles.

  The transducer can be used for an angular movement of 360°; it permits the detection of angular displacements as small as 1 angular second with proper stabilization of the supply voltage and of the ambient temperature.

  The output impedance of the convectron transducer varies between 50 and 200 ohms, depending upon the applied voltage; its time constant is about 0.1 sec. The noise level is 1 to 5 mV at a-c operation and considerably less at d-c operation. Random air currents striking the tube and ambient-temperature variation are likely to cause errors. The power requirements are less than 1 watt.

  The main advantages of the position convectron transducer are sensitivity, simplicity, and the absence of moving parts. Its disadvantage for a number of applications is the relatively large time lag. It should be noted that the transducer sensitivity depends upon the magnitude of the acceleration as well as upon its direction. If one of these tubes is displaced by an angle of 10° from the vertical and the output is balanced to zero, it will give a signal of 500 mV when the acceleration changes by 1 $g$.

For references concerning similar thermal displacement transducers (bolometer transducers), see P. M. Pflier, "Elektrische Messung mechanischer Grössen," 3d ed., no. 22, p. 235, Springer Verlag, Berlin, 1948. An application of the thermal transducer for the measurement of sound intensity (the oscillatory movement of the air cools an electrically heated wire) is described by W. S. Tucker and E. T. Paris, *Phil. Trans. Roy. Soc.* (*London*), (**A**) **221**, 389 (1920–1921).

### 1-25. Piezoelectric Transducers

If a force is applied to a solid crystalline dielectric, Fig. (1-2)58, it will produce stress within the crystal and a deformation of the crystal lattice. In certain crystals with asymmetrical charge distributions (e.g., quartz), the lattice deformation is, in effect, a relative displacement of the positive and negative charges within the lattice. The displacement of the internal charges will produce equal external charges of opposite polarity on opposite sides of the crystal (piezo-electric effect). The charges can be measured by applying electrodes to the surfaces and measuring the potential difference between them.

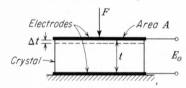

FIG. (1-2)58. Piezoelectric effect, schematic diagram.

For general literature on the piezoelectric effect, see W. P. Mason, "Piezoelectric Crystals and Their Application to Ultrasonics," D. Van Nostrand Company, Inc., 2d ed., Princeton, N.J., 1954; W. G. Cady, "Piezoelectricity," McGraw-Hill Book Company, Inc., New York, 1946; W. Voigt, "Lehrbuch der Kristallphysik," Teubner Verlagsgesellschaft m.b.H., Leipzig, 1928.

The magnitude and polarity of the induced surface charges are proportional to the magnitude and direction of the applied force $F$.

$$Q = dF \qquad (1)$$

where $d$ is a constant of the crystal (piezoelectric constant, see Table 5). The force $F$ causes a thickness variation of the crystal, i.e., a displacement $\Delta t$. The induced charge, written in terms of the displacement $\Delta t$, is

$$Q = d\,\frac{aY}{t}\,\Delta t \qquad (2)$$

where $a$ is the area of the crystal, $t$ its thickness, and $Y$ is Young's modulus.

$$Y = \frac{\text{stress}}{\text{strain}} = \frac{Ft}{a\,\Delta t}$$

The charge at the electrodes gives rise to a voltage $E_o = Q/C$, where $C$ is the capacitance between the electrodes. Since $C = \epsilon a/t$ ($\epsilon$, dielectric constant), the output *voltage* is

$$E_o = \frac{dt}{\epsilon a}\,F = g\,\frac{t}{a}F = gtP \tag{3}$$

where $g$ is a constant of the crystal (values for $g$ are given in Table 5), and $P$ is the pressure applied to the piezoelectric crystal.

The piezoelectric element is usually a slab, or "cut," taken from a

TABLE 5. PIEZOELECTRIC CONSTANTS

| Material | Orientation | Charge sensitivity $d$, coulomb/m² newton/m² | Voltage sensitivity $g$, volt/m newton/m² |
|---|---|---|---|
| Quartz | X cut; length along Y length longitudinal | $2.25 \times 10^{-12}$ | 0.055 |
| | X cut; thickness longitudinal | $-2.04$ | $-0.050$ |
| | Y cut; thickness shear | 4.4 | 0.108 |
| Rochelle salt | X cut 45°; length longitudinal | 435. | 0.098 |
| | Y cut 45°; length longitudinal | $-78.4$ | $-0.29$ |
| Ammonium dihydrogen phosphate | Z cut 0°; face shear | 48. | 0.354 |
| | Z cut 45°; length longitudinal | 24. | 0.177 |
| Commercial barium titanate ceramics | ‖ to polarization ⊥ to polarization | 130 to 160 $-56.$ | 0.0106 0.0042 to 0.0053 |

crystal. The magnitude and direction of the piezoelectric effect in the cut depend upon its orientation with respect to the crystal axes. Cuts are identified by the direction of the perpendicular drawn to the largest face of the cut; if this direction falls into the direction of the X axis of a crystal, the element is called X cut. Other more complex cuts are identified by their angular orientation with respect to the other two crystal axes.[1] The analytical evaluation of the properties of

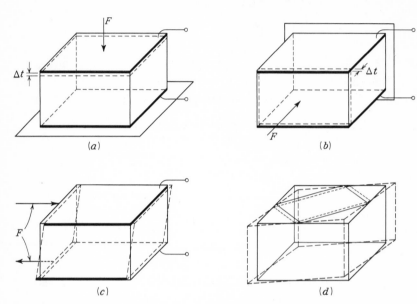

Fig. (1-2)59. Modes of operation of piezoelectric transducers: (a) thickness or longitudinal compression; (b) transversal compression; (c) thickness shear action; (d) face shear action (leading to length expansion) ("*Piezotronic Technical Data*" *from Brush Electronics Corp., Cleveland,; by permission*).

crystal cuts is complicated because of the mechanical and electrical anisotropy of the crystals and because of the variation of the crystal constants with varying mechanical load.

Piezoelectric elements made from barium titanate ceramics are free from limitations imposed by the crystal structure. They can be molded in different sizes and shapes, and the electrical axis can be "built in" in the production process.

MODES OF OPERATION, SINGLE PLATES. In single piezoelectric slabs or plates, four different modes of operation can be distinguished, depending upon the material and upon the crystallographic orientation of the plate. These modes are shown in Fig. (1-2)59.

[1] See examples for quartz in Mason, *op. cit.*, p. 83, fig. 6.3.

An electric output can arise either in response to a compression-expansion motion ((1-2)59 *a* and *b*) or in response to a shearing motion ((1-2)59*c* and *d*). Depending upon the material, a compression of the plate (*a* and *b*) can cause a diminution of the plate volume, or it can cause an expansion of the plate in other directions (transverse expansion), so that the plate volume remains constant. Most crystals exhibit a certain transverse expansion and a small net-volume change. Internal mechanical coupling within the crystal can cause the simultaneous response to compression-expansion and shear action. By the appropriate orientation of the cut and application of

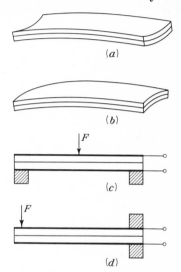

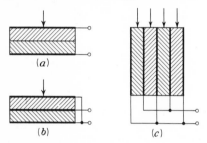

Fig. (1-2)60. Stacked piezoelectric crystals: (*a*) series connection; (*b*) and (*c*) parallel connections.

Fig. (1-2)61. Piezoelectric crystals combined to form "bimorph benders."

the electrodes, crystals can also be used for flexing or torsional movements.

The piezoelectric effect is reversible; the application of an electric field to a crystal causes a mechanical displacement of the crystal.

*Multiple arrangements* (stacks) of piezoelectric elements operating in the compression-expansion mode are shown in Fig. (1-2)60. The elements may be connected in series, leading to a higher output voltage for the same force, or in parallel, resulting in lower output impedance than a single element. Crystal stacks can also be arranged so that the response of neighboring crystals to an external vibration or an unwanted force is electrically compensated.

The single-plate piezoelectric transducer as shown in Fig. (1-2)59 or the multiple arrangements in Fig. (1-2)60 are useful for large driving forces but small displacements (order of microinches). If smaller forces or larger displacements are to be converted into electric signals, the piezoelectric systems illustrated in Figs. (1-2)61 and (1-2)62 are more suitable.

BIMORPHS, BENDERS. The elements shown in Fig. (1-2)61 (Bimorphs, trade name of the Brush Electronics Company) consist of two transverse expanding plates cemented together in such a manner that one plate contracts and the other expands. If a voltage is applied to such an element and if the element is permitted to move

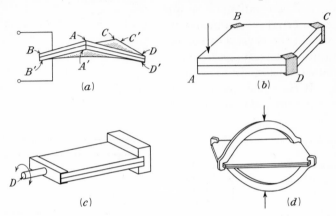

FIG. (1-2)62. Piezoelectric crystals combined to form "bimorph twisters."

freely, it will bend as shown (greatly exaggerated) in part *a* if it is made from a crystal, or in the form of part *b* if made from a piezoelectric ceramic (bimorph bender).

The bimorph bender can be used for input transducers; a bending of the element causes the development of a potential between the electrodes. Examples of such systems are shown in part *c* (end-supported, center-driven mounting) and part *d* (cantilever mounting). Systems of this type are frequently used in phonograph pickup cartridges.

TWISTERS. In Fig. (1-2)62*a* two face-shear plates are cemented together so that their expanding diagonals are perpendicular. If a voltage is applied to both plates and the system is permitted to move freely, it will assume a shape as shown, greatly exaggerated, in the figure (bimorph twister). This motion comes about because the diagonal $AC$ contracts while the diagonal $A'C'$ expands; similarly $BD$ expands and $B'D'$ contracts.

Mechanical twisting of these elements will cause a voltage between the electrode terminals. An application for input transducers is shown in Fig. (1-2)62*b*. The twister is held fixed at the three corners

*B, C,* and *D.* A displacement of the point *A* in the direction of the arrow will cause a twisting motion of the element and will give rise to an output voltage. Another application is shown in Fig. (1-2)62c, where a torsional motion of the drive shaft *D* causes a twisting motion of the bimorph and induces an output signal. A third application, illustrated in Fig. (1-2)62d, shows the conversion of a movement into the twisting of the piezoelectric element and, in further sequence, into an electric output signal.

The double curvature in bimorphs, both of the bender and of the twister type, requires special precaution in mounting the element. An ideal mounting should permit the desired bending or twisting motion without interfering with the normal flexure of the element. This is generally difficult; a practical approach is to hold bimorphs cemented or clamped between rubber pads.

The mechanical resonance frequency of bmiorphs is considerably lower than that of the single or stacked piezoelectric elements.

MATERIALS. Twenty of the thirty-two crystallographic classes exhibit piezoelectric properties, but only a few materials are practical for piezoelectric input transducers, primarily quartz, rochelle salt, ammonium dihydrogen phosphate (ADP), and ceramics made with barium titanate. Materials used less often are tourmaline, ethylene diamine tartrate, dipotassium tartrate, potassium dihydrogen phosphate, and lithium sulfate.

*Quartz* ($SiO_2$) is mechanically and thermally the most stable among the piezoelectric materials. Its internal electric losses are small, its volume resistivity is higher than $10^{14}$ ohm-cm. Because of its relatively small piezoelectric effects, its application is restricted to such uses where high tensile strength, high mechanical stability, or operation at elevated temperature is essential (the material can be operated safely up to 550°C). Quartz plates are applicable primarily for thickness and transverse compression-expansion operation, Fig. (1-2)59a and *b.*

For the technical preparation and testing of quartz elements, see R. A. Heising, "Quartz Crystals for Electrical Circuits," D. Van Nostrand Company, Inc., Princeton, N.J., 1946.

*Rochelle salt* ($NaKC_4H_4O_6 \cdot 4H_2O$) is stable at room temperature between 35 and 85 per cent relative humidity. At higher humidity it will deliquesce; at lower humidity it will dehydrate. Coating with wax is recommended to retard the humidity effects. At a temperature above 55°C the crystal will disintegrate.

The dielectric constant in the direction of the X axis of the crystal varies, within wide limits, with the temperature; also the resonance frequency of the crystal changes with the temperature.

The X cut of the rochelle salt crystal is primarily applicable for face-shear motion. Crystals can also be cut for transverse (longitudinal) compression-expansion, and the cuts can be combined to form bender and twister bimorph systems.

The volume resistivity is of the order of $10^{12}$ ohm-cm or more. The mechanical strength is considerably lower than that of quartz.

*Ammonium dihydrogen phosphate* ("ADP," $NH_4H_2PO_4$) has a higher thermal stability than rochelle salt; it is stable up to a temperature of 180°C, but is used, for technical reasons, with an upper limit of about 125°C. It operates satisfactorily in a range from zero to 94 per cent relative humidity if condensation of water on the surface (causing surface leakage) is avoided. The volume resistivity is considerably lower than that of the other substances described above; it is about $10^{10}$ ohm-cm and decreases strongly with increasing temperature and content of impurities.

The material (primarily the Z cut) is applicable for face-shear and longitudinal expansion operation and for twister bimorphs. The sensitivity is high. A crystal 1 cm thick furnishes for a load of 1 dyne/cm² an output voltage of $1.78 \times 10^{-4}$ volt.

*Barium titanate ceramics* ($BaTiO_3$). The single barium titanate crystal is ferroelectric, i.e., if exposed to an electric field it becomes polarized and maintains its polarization after the field is removed. Application of a field in the reverse direction causes a reverse polarization. The process follows a hysteresis curve.

In the ceramic polycrystalline form, the individual crystals are randomly oriented. By the application of a polarizing electric field (of order of 1,000 to several thousand volts per millimeter for about 1 hr at room temperature), the ferroelectric domains are lined up, and the material becomes piezoelectric. The material loses some of the piezoelectric activity in the first few days and remains fairly constant afterwards. The aging process can be accelerated by subjecting the material to higher temperature (80°C). The addition of about 4 per cent lead titanate increases the long-time stability.

Barium titanates are applicable for thickness compression-expansion and for transverse compression-expansion operation, Figs. (1-2)59a and b; they can be combined to form bender bimorphs, but they do not operate in shear modes and, consequently, they cannot be combined to form twister bimorphs.

The thickness compression and the transverse compression effects are opposite in sign but not equal in magnitude. A volume change produced by subjecting the transducer to pressure from all sides (hydrostatic pressure) causes an output voltage.

The voltage-output versus strain-input characteristic is not linear, except over a small operating range. A typical characteristic for a cylindrical plate loaded axially[1] is shown in Fig. (1-2)63. The linearity may be improved by prestressing of the transducer and operating in the central part of the characteristic.

SENSITIVITY. The piezoelectric coefficient $d$, or *charge* sensitivity, shown in Eqs. (1) and (2), varies between 1.4 and 6.3 × 10$^{-10}$ coulomb/newton. The charge sensitivity diminishes as the size of the element decreases.[2] Since the dielectric constant is very high, about 1,200 to 1,700 at room temperature, the coefficient $g$, or *voltage* sensitivity, shown in Eq. (3), is relatively small. Values reported range from 1.2 × 10$^{-2}$ to 6 × 10$^{-2}$ (volt/m)/(newton/m$^2$.) The piezoelectric constants can vary considerably with the temperature, the thermal history ("temperature hysteresis"), and the

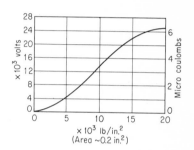

FIG. (1-2)63. Transfer characteristic of bariumtitanate transducer.

composition. In particular, addition of lead titanate ($PbTiO_3$) and calcium titanate ($CaTiO_3$) changes the piezoelectric constants and their variation with temperature, the dielectric constant and its variation with temperature, and the mechanical properties, i.e., Young's modulus and the resonance frequency of the piezoelectric elements. The piezoelectric activity of barium titanate ceases at a temperature above about 125°C (curie point). Crystallographic transitions and variations of the piezoelectric behavior also occur at about −10°C and −70°C. The variation occurring at −10°C is less pronounced, depending upon the composition of the ceramic body; with certain compositions it can be made to disappear completely.

The voltage sensitivity of barium titanate is less than that of rochelle salt and ammonium dihydrogen phosphate, but its temperature range is wider, its mechanical strength higher, and its insulation resistance is higher (10$^{13}$ ohm-cm). Compared with quartz, it has the advantage of lower cost. It can be made in various forms and shapes

[1] From E. H. Mullins, *Proc. Symposium on Barium Titanate Accelerometers, Natl. Bur. Standards Rept.* 2654, August, 1953.

[2] T. A. Perls, *Proc. Symposium on Barium Titanate Transducers, Natl. Bur. Standards Rept.* 2654, p. 51, 1953.

which would be difficult or impossible to cut from crystals, and the direction of the mechanical or electrical axis can be selected by polarization in the desired plane. Furthermore, the relatively large capacitance of the barium titanate piezoelectric elements can be considered an advantage. Mullins[1] has subjected barium titanate cylinders to large compressive forces and reports output voltages up to 60,000 volts just prior to crushing.

FREQUENCY RESPONSE (ELECTRIC). The equivalent circuit of a piezo-electric input transducer is shown in Fig. (1-2)64. $C_c$ is the capacitance of the piezoelectric element. $R_c$ is the leakage resistance of the piezoelectric element; it is usually high, of the order of $10^8$ to $10^{10}$ ohms. The resistance between the terminals is, in general, determined by the load resistance $R_L$ (order of $10^6$ to $10^7$ ohms). $C_L$ is the capacitance of the subsequent stage (load) plus that of the connecting cables.

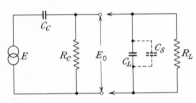

FIG. (1-2)64. Equivalent circuit of a piezoelectric transducer and electric load.

At medium and high frequencies, the voltage $E_o$ across the load is determined primarily by the capacitances $C_c$ and $C_L$. These capacitors form a voltage divider, the voltage $E_o$ is $EC_c/(C_c + C_L)$, independent of the frequency. If the output voltage is large, it can be reduced by an increase of $C_L$, i.e., by the parallel connection of a shunt capacitor $C_s$. At low frequencies, the voltage $E_o$ across the load is determined primarily by the reactance of $C_c$ and the impedance of the parallel combination $C_L$ and $R_L$. The voltage $E_o$ is then dependent upon the frequency and decreases with decreasing frequency. The low-frequency response can be improved by increasing $C_c$ (using a transducer with high capacitance) or by increasing the load resistance $R_L$. For a steady-state mechanical force the transducer furnishes no output voltage.

A means to overcome this difficulty and to measure steady-state loads with piezoelectric elements has been devised by Perls and Kissinger.[2] A crystal is driven by an electric oscillator at its resonance frequency, and the applied voltage (or the current) is measured. If mechanical stress is applied to the crystal, its resonance frequency or its power absorption will change. In the linear part of the operation characteristic of this device, the output varies by 4 volts for a

[1] Mullins, *loc. cit.*

[2] T. A. Perls and C. W. Kissinger, *Natl. Bur. Standards Rept.* 2390a, p. 12, June, 1955.

change of the applied steady-state load of 1 kg.  The method has also been used by the same authors for the measurement of steady-state air pressure surrounding a piezoelectric transducer.

The high-frequency limit is imposed by the mechanical resonance of the piezoelectric element plus associated mountings.  The frequency response of piezoelectric transducers may have spurious peaks.  Some are due to different modes of oscillations in the crystal, others to vibrations of the transducer mounting, resonances in attachments, cables, cooling tubes, etc.  Such resonances are usually of a lower frequency and can be eliminated by careful mounting.

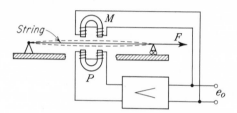

Fig.  (1-2)65.  Oscillating-string  transducer, schematic diagram.

The piezoelectric transducer is simple and rugged, suitable for many applications where sufficient mechanical force is available. The phonograph crystal cartridge is, at times, an excellent means for experimental work in the frequency range from 20 to about 20,000 cps;  the output voltage from such cartridges can be as high as several volts.  The piezoelectric transducer has a high output impedance and requires a high input impedance of the subsequent stage.  Its greatest disadvantage is its lack of response to steady-state ("d-c") displacements or forces.

**1-26. Miscellaneous Transducer Systems**

*a. Vibrating-string Transducer.*  The natural frequency of a string of the length $L$ and the cross-sectional area $a$ held between two fixed suspensions, Fig. (1-2)65, is

$$f = \frac{1}{2L}\sqrt{\frac{F}{as}} = \frac{1}{2L}\sqrt{\frac{\sigma}{s}} \tag{1}$$

where $F$ is the force which holds the string taut and causes a tensile stress $\sigma \,(= F/a)$ in the wire, and $s$ is the density of the wire material. The force $F$ also causes the wire of the original (unstretched) length

$L$ to be lengthened by the amount $\Delta L$, and since $\Delta L/L = \sigma/Y$ ($Y$, Young's modulus of elasticity), the natural frequency can be expressed by

$$f = \frac{1}{2L}\sqrt{\frac{Y}{s}\frac{\Delta L}{L}} \tag{2}$$

Any change of the length or of the force or of the stress in the wire causes a corresponding change of the natural oscillation frequency of the wire.

The mechanical oscillation of the wire is converted into an electric signal usually by means of an inductive pickup system $P$ located near the oscillating string. The (steel) wire, in vibrating, varies the magnetic flux in the air gap of the pickup system; the output signal is amplified and fed back into an electromagnet $M$ to keep the string excited in its natural frequency.

The transducer has been used for a range of stress in the string between 20 and 60 kg/mm². If the stress becomes too low (i.e., lower than 15 kg/mm²) the wire generally will not produce a pure sinusoidal oscillation; if the stress is too high, it will exceed the limit of proportionality. The maximum displacements that can be measured are of the order of one-thousandth of the wire length. The minimum displacement that can be observed can be very small, since frequency deviations can be measured with a high amount of sensitivity. It has been claimed that displacements as low as $0.1\,\mu$ can be detected.

The error of the method is, in general, between 0.1 and 1 per cent. The frequency variation is a nonlinear function of the displacement or the force as evident from Eqs. (1) and (2). The process to be investigated must be long compared to the oscillation period of the wire; variations occurring in $\frac{1}{100}$ of a second have been measured.[1]

The natural frequency of the wire varies strongly with temperature. A variation of temperature from 0 to 40°C causes the same variation in the output as does a change in stress by 10 kg/mm². Thermal compensation by mounting the wire on a base with the same thermal expansion coefficient as the wire is frequently required. The length of the string is in general between 20 and 200 mm; if it is too short, end conditions become too important and the relationships of Eqs. (1) and (2) are no longer valid. A ratio of length to wire diameter of more than 250 is recommended.

The oscillating-string transducer is primarily applicable for large forces. If thermally protected, its long-time operating stability can

---

[1] H. Moser, *Bull. schweiz. elektrotech. Ver.*, **25**, 689 (1934).

be very good.  The fact that the output appears in the form of a variable frequency signal can be advantageous in digital systems.

For literature references, see O. Schaefer, *Z. Ver. deut. Ing.*, **63**, 1008 (1919); R. S. Jerret, *J. Sci. Instr.*, **22**, 29 (1945).

*b. Electronic Displacement Transducers.*  The plate current in a space-charge-limited vacuum tube is a function of the electrode geometry.  In a diode with a plane parallel anode and cathode of an area $a$ spaced by a distance $d$, the electron current is

$$I = 2.34 \times 10^{-6} \times \frac{a}{d^2} E^{\frac{3}{2}}$$

$E$ is the applied plate voltage.  The plate current can be changed by variation of the separation $d$.

A diode transducer based upon this principle is described by Day[1] and is shown schematically in Fig. (1-2)66.  It consists of an evacuated tube, closed at one end with a flexible membrane;  the cathode

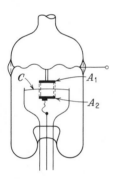

Fig. (1-2)66. Moving-anode displacement transducer used for the measurement of pressure, schematic diagram.

Fig.     (1-2)67. Moving - anode transducer, schematic diagram (*mechano-electronic transducer tube, RCA type* 5734).

assembly $C$ is held in a fixed position, while two anodes $A_1$ and $A_2$ are held by a stem supported in the center of the membrane.  A variation of the diaphragm position increases the distance from the cathode to one anode and decreases that to the other anode.  The system is used in a bridge arrangement;  the transfer characteristic (voltage in the

[1] G. Day, *J. Sci. Instr.*, **26**, 372 (1949).

bridge diagonal versus negative pressure applied to the outside of the membrane) is linear within about 2 per cent. The system is temperature sensitive, but the influence of the temperature upon the transducer output can be nearly compensated by a thermistor circuit.

A similar construction is described by W. Ramberg, *J. Research Natl. Bur. Standards*, **37**, 391 (1946), for use as an accelerometer. The acceleration causes the movement of two plates within the tubes. See also R. Gunn, *Rev. Sci. Instr.*, **11**, 204 (1940), and *J. Appl. Mechanics*, **7** (1940), and also U.S. Patent 2,155,420, Feb. 27, 1940, and J. Rothstein, U.S. Patent 2,389,935, Nov. 27, 1945.

Higher sensitivities can be obtained with a triode system, as shown in Fig. (1-2)67 (Mechano-electronic Transducer, RCA Tube 5734). The cathode and grid assembly are held in a fixed position within a vacuum-tight envelope, the anode is supported by a rod which extends through the center of a thin metal diaphragm sealed to the tube envelope. An angular displacement of this rod leads to a variation of the plate current. The transfer characteristic is linear within about 2 per cent. The maximum permissible displacement of the rod is $\pm 0.5°$ about the zero position. For this displacement, a torque of 13.3 g-cm is required. The moment of inertia of the moving system is 3.4 mg-cm$^2$. The frequency response is limited by the mechanical resonance of the part of the plate shaft within the tube, which is about 12,000 cps. The tube is generally operated with the grid connected to the cathode and with a plate supply voltage of 300 volts applied through a 75,000-ohm load resistor. A maximum displacement of the anode rod by 0.5° results in a variation of the output by $\pm 20$ volts. Thermal drift and some low-frequency instability at a signal level of 1 mV, as well as variations of the output by 1 per cent of the maximum signal for a variation of the heater voltage by 1 per cent have been observed by Talbot.[1]

*c. Ionization Transducer.* A glass tube filled with gas at a pressure of about 10 mm Hg and containing two electrodes is brought into an electric high-frequency field between the plates $P_1$ and $P_2$ of a capacitor, as shown in Fig. (1-2)68a. If the field is sufficiently high, a glow discharge will arise in the tube. The two electrodes $A$ and $B$ act as probes in the discharge; their potential is determined by the space potential of the plasma surrounding each electrode and by the rf potential induced by their capacitive coupling to the plates $P_1$ and $P_2$. In the symmetry position the net charges of both electrodes are equal, so that their potential difference is zero. Outside of the symmetry position, the charges are different for each electrode and give

[1] S. A. Talbot et al., *Rev. Sci. Instr.*, **22**, 233 (1951).

rise to a d-c potential difference. The transfer characteristic, i.e., the output voltage $E_o$ versus the displacement $x$, is shown in Fig. (1-2)68$b$. The potential difference can reach values of more than 100 volts. The sensitivity of the arrangement, $\Delta E_o/\Delta x$, can reach values up to several volts per micron of displacement. The phenomenon can be observed from very low excitation frequencies (60 cps) up to the microwave region. For technical reasons, operation between 0.1 and 10 Mc is recommended. Accurate frequency stability is not required for the operation of the transducer. The lifetime of the tube can be

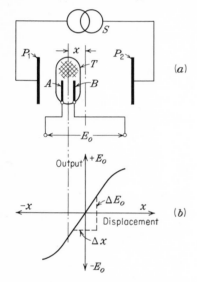

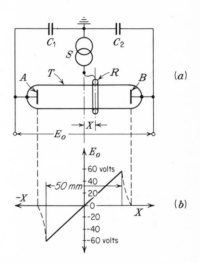

FIG. (1-2)68. Ionization transducer: (a) basic circuit; (b) transfer characteristic [*figures* (1-2)68 *and* (1-2)69 *from K. S. Lion, Rev. Sci. Instr.*, **27**, 222 (1956); *by permission*].

FIG. (1-2)69. Ionization transducer system for large movements: (a) schematic diagram; (b) transfer characteristic.

considerably increased by arranging the plates $P_1$ and $P_2$ in such a way that the glow discharge is largely confined to that part of the tube opposite the electrodes. This is the cross-hatched part of the tube in Fig. (1-2)68$a$. The glow surrounding the electrodes $A$ and $B$ is then reduced, and cathode sputtering is eliminated.

The arrangement of Fig. (1-2)68 is useful for the conversion of displacements up to about 1 mm. Figure (1-2)69 shows a transducer system for large movements up to several inches. The tube $T$ is filled with gas under reduced pressure. One side of a high-frequency source $S$ is connected through capacitors $C_1$ and $C_2$ to the internal

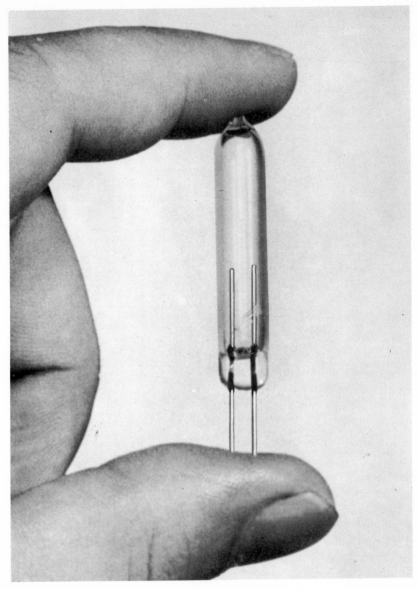

Fɪɢ. (1-2)70. Ionization transducer tube (*courtesy of The Decker Corporation, Bala-Cynwyd, Pa*).

electrodes $A$ and $B$; the other side of the source $S$ is connected to a ring electrode $R$ which externally surrounds the tube. A discharge takes place between $A$ and $R$ as well as $B$ and $R$. Any asymmetry of this arrangement brought about by longitudinal shifting of the ring electrode causes a d-c voltage to arise between $A$ and $B$. The purpose of the capacitors $C_1$ and $C_2$ is to prevent shorting of this d-c voltage. The output voltage $E_o$ arising between the internal electrodes $A$ and $B$ varies linearly with the displacement of the external ring electrode over a wide range. The sensitivity is in general between 1 and 10 volts/mm, which is considerably lower than in the transducer system of Fig (1-2)68. Also this system is relatively insensitive to variation of the supply source; a change of 10 per cent of the frequency and also a variation of 10 per cent of the supply voltage are barely noticeable in the output.

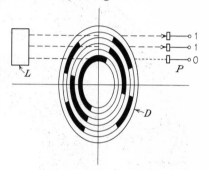

FIG. (1-2)71. Mechanic-optical transducer, shaft-position encoder with digital read out; $L$, light source; $D$, encoder disk; $P$, phototransducers.

Variation of the symmetry can also be obtained by capacitance-variation circuits. The output impedance of the device is of the order of 1 M$\Omega$; therefore, the succeeding stage must have a high input impedance. The system is applicable for mechanical displacements of a steady-state character and up to mechanical frequencies of several thousand cycles per second.[1]

*d. Mechanic-Optical Transducers.* Mechanical displacements can be converted into electric signals by optic-electrical means, for instance, by an arrangement consisting of an illuminated slit and a photoelectrical transducer; the moving object obscures a part of the slit and causes a variation of the light intensity reaching the photoelectrical transducer and hence a variation of the transducer output.

An arrangement of this type used primarily for the conversion of rotary displacement into a digital output ("shaft-position encoder") is shown schematically in Fig. (1-2)71. An optical system $L$ produces a number of parallel light beams which are directed upon corresponding photo transducers $P$. The light beams traverse an encoder

[1] K. S. Lion, *Rev. Sci. Instr.*, **27**, 222 (1956). A commercial ionization transducer tube (T-42, The Decker Corporation, Philadelphia) is shown in Fig. (1-2)70.

disk $D$ with transparent and opaque segments. Depending upon the position of the disk, some of the light beams will reach the photo transducers and cause an output signal; the result appears in binary digits (e.g., 1, 1, 0).

An encoder disk with 10 concentric arrays of segments, for 10-digit information, is shown in Fig. (1-2)72; a disk of this type gives a

FIG. (1-2)72.  Encoder disk for 10 digits.

different output for each of the $2^{10}$ possible configurations, i.e., it furnishes information of the angular-disk position with an accuracy of $360°/1,024$, or an angle of about $0.35°$. Systems of this type have been built with disks of 10 in. diameter having up to 17 concentric segments. An accuracy as high as 10 sec of arc has been obtained. Instantaneous reading of rotating disks is accomplished by pulse operation of the light source.[1]

[1] Information by courtesy of Wayne-George Corporation, Boston, Mass.

## 1-3. Transducers for Velocity or Acceleration

*a. Differentiating Systems.* Electric signals which are proportional to the velocity, the acceleration, or to a higher time derivative of a displacement are usually formed by electrical differentiation of the displacement transducer output, as shown in Fig. (1-3)1. A method,

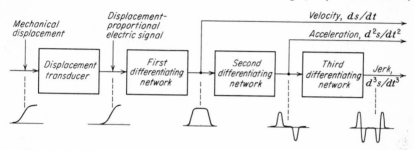

Fig. (1-3)1. Circuit for obtaining output signals proportional to velocity and acceleration from displacement transducers.

based upon this principle, which permits the measurement of acceleration and force from displacement-proportional signals (measurement of the force with which the extrinsic ocular muscles move the eyeball), is described by Lion and Powsner.[1]

Instead of starting out from an electric signal analogous to a movement or a force and forming the derivatives electrically, as shown in

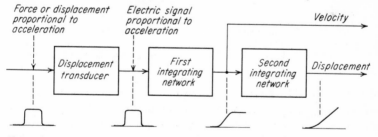

Fig. (1-3)2. Circuit for obtaining velocity and displacement proportional signals from acceleration–proportional input signals.

Fig. (1-3)1, one can also produce by mechanical means a displacement that is proportional to a velocity or an acceleration and convert this displacement into an electric signal. For instance, under the influence of an acceleration, a mass may exercise a force or cause a displacement which is proportional to the acceleration. The force or

[1] K. S. Lion and E. R. Powsner. *J. Appl. Physiology*, **4**, 276 (1951); see also R. J. Brockhurst and K. S. Lion, *Arch. Ophthalmol. (Chicago)*, **46**, 311 (1951).

the displacement can be converted into an electric signal with one of the displacement transducers described above. Such a signal can be converted into a velocity-proportional signal by the use of an integrating network, as shown in Fig. (1-3)2. In general, systems involving integrating networks are preferable to those illustrated in Fig. (1-3)1, since differentiating networks tend to increase the noise level.

*b. Induction Transducers.* The only transducer which converts the velocity of a movement directly into an electric signal is the

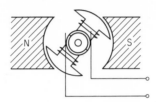

FIG. (1-3)3. Tachometer for d-c output with fixed magnet and movable armature.

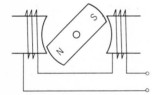 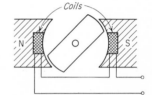

FIG. (1-3)4. Tachometer for a-c output with fixed armature and moving magnet.

FIG. (1-3)5. Tachometer for a-c output with fixed magnet and coils and moving yoke.

induction transducer described in 1-22*f*. The system is frequently used for the measurement of angular velocity, i.e., as a tachometer. Three different forms of such tachometers are commonly available. The first, shown in Fig. (1-3)3, is a d-c generator usually built with stationary magnet and rotating armature; the output from the armature is led to a collector. The magnitude of the output voltage is generally of the order of 10 mV/rpm. The system permits the use of d-c instruments for the measurement of speed and direction. In some applications, the collector noise and ripple voltages are likely to cause difficulties. The second system is shown in Fig. (1-3)4; it consists of a rotating magnet and a stationary coil and furnishes an a-c output. A third form, also for a-c output, is illustrated in Fig. (1-3)5; it has both a stationary magnet and a stationary coil but uses a rotating yoke of iron or other material with high magnetic permeability, which alternatingly increases and decreases the flux passing through the coil. An alternate form in which an iron body rotates in the gap between a stationary magnet and a stationary coil is shown by Pflier.[1]

The output voltage from all three systems is proportional to the rotation velocity, provided the load resistance connected to the terminals is very high. If the load current is not negligible, a magnetic field arises in the coil which can distort the primary (exciting) magnetic field and give rise to nonlinearity. Errors may arise from temperature variation. The residual flux and the coercive force of permanent magnets decrease in general at higher temperatures. The effect can be reduced by using electromagnets instead of permanent magnets and compound (shunt-type) windings. Measurement of the output *frequency* from a-c generators can furnish more accurate results; however, since most frequency-measuring systems demand constant voltage, means have to be provided to keep the output voltage from the tachometer constant, i.e., independent of the rotation velocity. This can be accomplished either with compound windings similar to those used in automobile generators or by a subsequent stage, e.g., a driven multivibrator. The most accurate results can be obtained by the use of digital systems, i.e., by counting the number of cycles during a measured time interval.

c. *Separate Measurement of $\Delta s$ and $\Delta t$.* If a body moves through a distance $\Delta s$ during a time interval $\Delta t$, a velocity-proportional signal can be obtained by producing electric signals proportional to $\Delta s$ and $\Delta t$ and forming the ratio $\Delta s/\Delta t = v$ by electrical means. The method is usually modified by keeping $\Delta s$ constant and measuring $\Delta t$ electrically, or vice versa. An example of a circuit used to determine the velocity of a bullet or an explosion wave is illustrated in Fig. (1-3)6. The two switches (or thin wires) $A$ and $B$ are a distance $\Delta s$ apart and are closed initially. Opening of the switch $A$ causes the appearance of a voltage between the output terminals (the voltage drop across $R_2$) which disappears at the opening of the switch $B$. The length of the voltage pulse is $\Delta t$ and is inversely proportional to the velocity.

Numerous variations of this method have been described, in particular in connection with repetitively acting devices such as in tachometers. Either one or several switches or commutators are used; either the pulse integral or the average pulse height is measured (e.g., in "point dwell meters" for the examination of ignition systems in

---

[1] P. M. Pflier, "Elektrische Messung mechanischer Grössen," 3d ed., p. 40, Springer Verlag, Berlin, 1948.

combustion engines) or the pulse time or the pulse frequency are measured.

In the capacitive tachometer a capacitor is alternately charged from a voltage source and discharged through a meter, as shown in

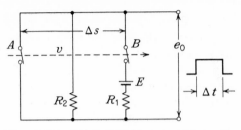

FIG. (1-3)6. Circuit furnishing an output pulse of a duration inversely proportional to the velocity of a bullet traveling the distance $\Delta s$ and opening, in succession, the switches $A$ and $B$.

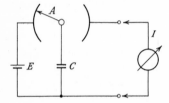

FIG. (1-3)7. Capacitive tachometer; the output current is proportional to the rotational velocity of the contact arm $A$ but independent of the direction of rotation.

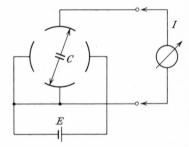

FIG. (1-3)8. Capacitive tachometer with direction-sensitive indication.

Fig. (1-3)7. The average discharge current $I$ is proportional to the rate of operation, $n$.

$$I = CEn$$

Another form of such a capacitive tachometer is illustrated in Fig. (1-3)8. This system, in contrast to that shown in Fig. (1-3)7, furnishes an output current which changes direction if the sense of rotation is inverted.

For general references on tachometers, see the book by Pflier, *op. cit.*, p. 244; also Ivan A. Greenwood, Jr., J. Vance Holdam, Jr., and Duncan MacRae, Jr., "Electronic Instruments," in M.I.T. Radiation Laboratory Series, vol. 21, sec. 4.5, and John F. Blackburn, "Components Handbook," in M.I.T. Radiation Laboratory Series, vol. 17, pp. 372ff, both McGraw-Hill Book Company, Inc., New York, 1948.

## 1-4. Liquid-level Transducers

A direct conversion of a liquid-level position into an electric signal is rarely used. In most cases the level position is detected indirectly, either by the mechanical movement of a float which is connected to a displacement transducer, by hydrostatic or pneumatic-pressure measurements, by optical or acoustic means, or by gamma-ray absorption (1-15).[1]

### 1-41. RESISTIVE METHODS

*a. Contact Systems.* The simplest direct-acting liquid-level transducer consists of one or more contacts at appropriate levels of the liquid; an electric circuit is closed through the liquid when it reaches the contact. If the resistivity of the liquid is too high for convenient current measurement or if the resistivity is variable, the level variation of the liquid may be made to change the length of a mercury column, so that the opening and closing of the circuit take place between the mercury and the contact points. A system of this type with 10 contacts (in 10 U tubes) has been described by Norman[2] and is schematically shown in Fig. (1-4)1.

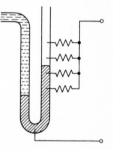

FIG. (1-4)1. Liquid-level transducer with multiple contacts.

The contacts can be connected to a circuit in such a way that the variation of the mercury column causes a stepwise variation of a resistance. The resistance variations can be made to be a linear function of the length of the mercury column or any other desired function. A function in which the resistance variation is proportional to the root of the mercury column displacement has sometimes been used for orifice-type flow meters; in these meters the flow velocity is proportional to the root of the pressure difference. Contact manometer systems with electrical temperature and pressure compensation (incorporating a second manometer and a resistance thermometer) are described by Lohmann and von Grundherr,[3] and by Geyger.[4] The contact systems have the disadvantage that the

[1] For a synopsis of such mostly mechanical systems, see *Bulletin* 1161, sec. 4, pp. 8 and 9, Minneapolis-Honeywell Regulator Company, Philadelphia, Pa.

[2] E. E. Norman, *Elec. World,* **120,** 635 (1943).

[3] H. Lohmann and F. von Grundherr, *Arch. tech. Messen,* V 1245-2, May, 1934.

[4] W. Geyger, *Arch. Elektrotech.,* **28,** 57 (1934).

contact points tend to deteriorate or corrode and that the indication is discontinuous.

*b. Electrolytic Resistance.* A continuous method to convert the level variation of a moderately conducting liquid (water, sea water) into electric signals is shown in Fig. (1-4)2. It consists of two wires A and B held at a constant distance a, for instance by means of insulating spacers C, and immersed in the liquid. The resistance between the wires varies in inverse proportion to the height h of the liquid level. Therefore, if a constant voltage is applied, the current through

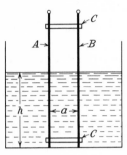

Fig. (1-4)2. Liquid-level transducer for use in conducting liquids (electrolytes).

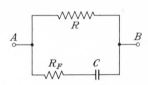

Fig. (1-4)3. Equivalent circuit of the electrolytic-level transducer shown in Fig. (1-4)2. $R$, resistance of the electrolyte; $C$, capacitance between the electrodes; $R_F$, equivalent loss resistance of the capacitance $C$.

the liquid is proportional to the level h. The current sensitivity of the method, $\Delta I / \Delta h$ (variation of the current $I$ caused by a variation of the liquid level h), decreases approximately hyperbolically with the distance a between the wires. One can, therefore, increase the current sensitivity by bringing the two wires close together, but such a procedure will increase the difficulty of keeping the distance between the wires constant, and a small variation in distance will produce a large error.

Electrolytic polarization makes it necessary, in most cases, to use alternating current for the determination of the resistance variation. The impedance between the terminals of A and B depends upon the resistance as well as the capacitance between both wires and can be represented by the equivalent circuit of Fig. (1-4)3. The magnitude of the capacitance C can be considerable.[1]

Capillary forces between the wires and the liquid tend to cause errors. Under static conditions (no movement of the liquid surface) the liquid rises on the wire by an amount which depends upon the wire diameter and the surface tension of the liquid and which is of

---

[1] W. Oesterlin, *Arch. tech. Messen*, V 1123-12, November, 1952.

the order of a millimeter if the wires are clean and immersed in clean water. Contamination of the wires can cause considerable variations of this value. Under dynamic conditions, i.e., if the water rises or falls, the water line around the wires can remain behind the true water level by several millimeters. Where possible, the admixture of a wetting agent to the liquid, to reduce surface tension, is recommended.

Variations of the resistivity of the liquid, brought about by a change of concentration of dissolved matter or a change of temperature, will cause errors. Two methods have been applied to reduce such errors. One method uses auxiliary contact points arranged at fixed (known) levels $h_1$, $h_2$, etc. As the liquid rises, its surface makes electrical contact with one of these points. This occurrence is recorded by a mark on the continuous record of the liquid level from the electrodes $A$ and $B$, and the continuous record can thus be calibrated. Another method uses an auxiliary electrode, as shown in Fig. (1-4)4. Three electrodes $A$, $B$, and $C$ are immersed in the liquid. A change of the liquid level causes a variation of the resistance between the electrodes $A$ and $B$ but not of that between $B$ and $C$. The resistance ratio, measured in a bridge or with a ratio meter, is practically independent of the conductivity of the liquid.

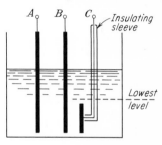

Fig. (1-4)4. Liquid-level transducer with compensation for conductivity variation of the electrolyte.

An instrument of this type which measures the water level on a weir, and thus the flow rate, has been described.[1] Ware[2] describes a liquid-level indicator which is based upon the resistance variation between the liquid and a servosystem-controlled contact bar.

## 1-42. Inductive Method

An inductive method for the detection of the level of liquid sodium in stainless-steel tubes has been described by Pulsford.[3] The arrangement, Fig. (1-4)5, consists of two coils $L_1$ and $L_2$ of about 1,000 turns wound around a steel tube $T$ containing the liquid sodium. The inductance of each coil is about 250 $\mu$H. One coil (search coil) can be set at a predetermined level. The inductance of the search coil

[1] Anon., *The Electrician (London)*, **89**, 209 (1922).

[2] L. A. Ware, *Electronics*, **13**, 23 (March, 1940).

[3] E. W. Pulsford, *J. Sci. Instr.*, **32**, 362 (1955).

changes rapidly as the liquid sodium moves into the plane of the coil. The method is successful because the tube material is but feebly magnetic, and the liquid metal is a good electrical conductor. The transfer characteristic is nonlinear. A similar system for the detection of conductive liquids (acids, etc.) is described by Broadhurst.[1]

A different construction for the conversion of the level variation of a mercury column into an electric signal is shown in Fig. (1-4)6.

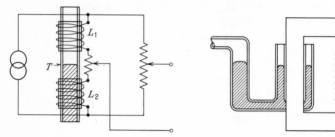

FIG. (1-4)5. Inductive-level transducer for liquid metals.

FIG. (1-4)6. Alternative arrangement of a level transducer for liquid metal (mercury).

The mercury column surrounds the iron core and forms the secondary winding of a transformer. Depending upon the level of the mercury, the resistance of this winding (and, thus, the inductance, the power consumption, or the current on the primary side of the transformer) varies and furnishes a signal which is a function of the mercury level.[2]

## 1-43. CAPACITIVE METHODS

*a. Direct Capacitance.* Direct capacitive methods for liquid-level detection are applicable primarily to nonconductive liquids, such as liquid gases, oil, or gasoline. The liquid forms the dielectric of a capacitor; its dielectric constant is higher than that of the gas phase above the liquid. Any rise of the liquid level increases the capacitance between the two electrodes. The electrodes are frequently arranged in the form of concentric cylinders, as shown in Fig. (1-4)7. At the lower end of the external cylinder are openings $B$ which permit hydraulic communication with the outside liquid. If these openings are sufficiently small, they will cause a mechanical damping of the surface variations of the liquid, which is sometimes desirable.

The capacitance of a cylinder is

$$C = \frac{\epsilon h}{2 \ln (r_2/r_1)} \tag{1}$$

[1] J. W. Broadhurst, *J. Sci. Instr.*, **21**, 108 (1944).
[2] G. Ruppel, *Arch. tech. Messen*, V 1244-1, June, 1933.

where $\epsilon$ is the dielectric constant of the medium between the cylinders, $h$ the length of the cylinders, and $r_1$ and $r_2$ the radii of the cylinder surfaces adjacent to the liquid. If the cylinder capacitor is only partly filled with a liquid, the capacitance of the arrangement shown in Fig. (1-4)7 is, for $h \gg r$ and $r \gg a$, in good approximation,

$$C = \frac{\epsilon' h' + \epsilon'' h''}{2 \ln\left[1 - (a/r)\right]} \qquad (2)$$

where $h'$ is the height of the liquid level, $h''$ the length of the cylinder above the liquid surface, as indicated in Fig. (1-4)7, $\epsilon'$ the dielectric constant of the liquid, $\epsilon''$ the dielectric constant of the gas above

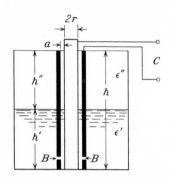

FIG. (1-4)7. Capacitive liquid-level transducer for dielectric liquids.

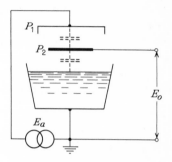

FIG. (1-4)8. Capacitive level transducer based upon the capacitance between the liquid surface and an auxiliary electrode $P_2$.

the liquid (usually $= 1$), $r$ the outside radius of the inner cylinder, and $a$ the distance between the inner and outer cylinders. The stray capacitance on both cylinder ends is usually in the order of several $10\mu\mu$F and must be added to $C$ in Eq. (2). The outer cylinder should be grounded.

Errors are likely to occur from temperature variations which cause a variation of the dielectric constant of the liquid. The error in liquefied gases may be as high as 10 per cent.

For references see W. E. Williams, Jr., and E. Maxwell, *Rev. Sci. Instr.*, **25**, 111 (1954); J. G. Dash and H. A. Bowse, *Phys. Rev.*, **82**, 1951 (1951).

*b. Electrode-to-Liquid Surface Capacitance.* If the conductivity or the dielectric constant of the liquid is sufficiently high (e.g., mercury, water), the capacitance between the liquid surface and a reference electrode above the surface can be used for liquid-level determination. A system of this type has been described by Hazen[1] and is

[1] H. L. Hazen, *Elec. Eng.*, **56**, 237 (1937).

shown schematically in Fig. (1-4)8. It consists of two plates $P_1$ and $P_2$ kept at a fixed distance and electrically insulated from each other. An a-c voltage $E_a$ (700 volts, 60 cps) is applied between the electrode $P_1$ and the water. The electrode $P_2$ assumes an a-c potential between zero and that of $P_1$; this potential varies with the level of the water surface.

The system can be used in two different ways: With stationary electrodes, the output voltage serves as a measure of the liquid level; the output voltage decreases with increased water level. The transfer characteristic (output voltage versus water level) is nonlinear. Another method of operation consists in the use of a servosystem which moves the two electrodes $P_1$ and $P_2$ up and down and keeps the distance between $P_2$ and the water surface constant. The latter method furnishes more accurate results. In the practical embodiment, the minimum distance between $P_2$ and the water surface is 2.5 cm. The instrument operates over a total range of water-level variation of 4.5 cm, and the error is in the order of 0.5 mm.

### 1-44. Thermal Method

This method has been applied primarily for the level determination of liquefied gases but is, in principle, applicable to other systems. A schematic diagram is shown in Fig. (1-4)9.

A fine wire $W$ extends vertically through a container which is

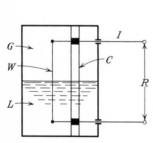

Fig. (1-4)9. Thermal liquid-level transducer.

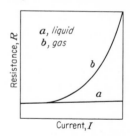

Fig. (1-4)10. Characteristics of the transducer shown in Fig. (1-4)9: (a) wire immersed in liquid; (b) in gas.

partly filled with a liquid. The wire is heated by a constant current. The heat transfer from the wire to the liquid is greater than that from the wire to the vapor. Thus, that part of the wire immersed in the liquid will be cooler and its resistance lower than the part extending in the vapor atmosphere. As the level of the liquid recedes, the average temperature of the wire, and thus its resistance, will increase.

The resistances of both the part of the wire in the liquid and that in the vapor atmosphere vary with the applied current, as shown in Fig. (1-4)10 (generalized characteristic, relative values). In the liquid (curve *a*), the resistance increases only slowly with increased current (only at a higher critical current value does the resistance increase sharply). In the vapor (curve *b*), the resistance rises with gradually increasing slope. The exact location of these curves varies with the medium and with the temperature.

The sensitivity increases with decreased wire diameter. Platinum wires (Wollaston wire, silver removed) of about 10 $\mu$ diameter or platinum ribbons are recommended. The calibration characteristic (current in the diagonal of a Wheatstone bridge versus liquid level) is nearly linear if the vapor is at the same temperature as the liquid.

A liquid-nitrogen-level indicator based upon this method is described by Maimoni.[1] The instrument permits the determination of the liquid level to within 0.1 in. and has a working range of 2 in.; changes in the liquid level over this range causes the current in the bridge diagonal to vary by 20 $\mu$A. The power dissipation is 3.4 mW/in. Wexler and Corak[2] describe an instrument with a very small wire which, when arranged in a horizontal position, permits the determination of the liquid level to better than 0.2 mm.[3]

## 1-5. Pressure Transducers

The field is divided into two sections, one for transducers operating at high pressures above atmospheric pressure and one for operation at pressures considerably lower than atmospheric pressure (vacuum gauges).

Direct-acting transducers for the high-pressure range are based either upon the variation of a resistance (1-51) or of a capacitance (1-52) or upon piezoelectric effects (1-53), electrokinetic effects (1-54), or electric gas discharges (1-55).

Direct-acting vacuum gauges are based upon either thermal conduction (Pirani gauge, 1-56) or upon ion currents in gases (Ionization Gauges, 1-57).

A great number of indirect-acting pressure transducers have been described in which the pressure to be measured causes the mechanical deformation of a membrane or of a Bourdon tube; the deformation

---

[1] A. Maimoni, *Rev. Sci. Instr.*, **27**, 1024 (1956).

[2] A. E. Wexler and W. S. Corak, *Rev. Sci. Instr.*, **22**, 941 (1951).

[3] See also E. H. Quinnel and A. H. Futch, *Rev. Sci. Instr.*, **21**, 400 (1950), and M. S. Fred and E. G. Rauh, *Rev. Sci. Instr.*, **21**, 258 (1950).

is then converted into electric signals by any one of the mechanical displacement transducers described above. Since these systems of conversion of pressure into electric signals represent merely combinations of elements, they are omitted. A thermoelectrical pressure transducer containing a boiling liquid is described by Goodyear.[1] The boiling temperature of the liquid varies with the pressure and is measured by a thermistor.

### 1-51. RESISTIVE PRESSURE TRANSDUCER

If a wire is subjected to an increase of external pressure exerted from all directions, Fig. (1-5)1, its electrical resistance will change.

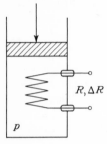

The resistance will usually decrease with pressure, although in some metals it increases, and in cesium it decreases first and increases at higher pressures. The effect has been extensively studied by Bridgman[2] and is caused by a distortion of the crystal lattice in the metal brought about by the external pressure.[3]

For many metals and over small ranges of pressure variations, the resistance changes linearly with a variation of pressure $\Delta p$.

FIG. (1-5)1. Bridgman-type pressure transducer, schematic diagram.

$$R = R_0(1 + b\,\Delta p) \tag{1}$$

where $R_0$ is the resistance at 1 atm pressure and $b$ the pressure coefficient of resistance. Average values of the pressure coefficient of resistance for various metals are given in Table 6.

The coefficient $b$ is largest for the alkali metals and for bismuth and antimony, but transducers made from such materials are not practical, and mercury and manganin are preferred. Mercury has the advantage that it can be produced with uniform purity, but it undergoes a phase transition near room temperature in the pressure range between 6,500 and 7,600 $kg/cm^2$; in this stage the pressure coefficient of resistance changes slightly (see Table 6).

The pressure coefficient of resistance does not change much with

[1] R. S. Goodyear, *Elec. Mfg.*, p. 90, October, 1956.

[2] P. W. Bridgman, "The Physics of High Pressure," The Macmillan Company, New York, 1931.

[3] N. H. Frank, *Phys. Rev.*, **47**, 282 (1935), and J. C. Slater, "Introduction to Chemical Physics," chap. 29, McGraw-Hill Book Company, Inc., New York, 1939.

temperature. Bridgman[1] has shown that for many metals at a pressure of 12,000 kg/cm$^2$ and in the temperature range between 0 and 100°C the change of $b$ with temperature is of the order of a few per cent; in some metals it is even less. However, the transducer must still be used at a constant temperature, since the resistance of the probe changes strongly with temperature.

The transducer is suitable for measuring pressures in the range from zero to above ten thousand kilograms per square centimeter.

TABLE 6. AVERAGE PRESSURE COEFFICIENTS OF RESISTANCE $b$ FOR DIFFERENT METALS IN THE RANGE 0–12,000 kg/cm$^2$ AT ROOM TEMPERATURE (25°)*

| Metal | Pressure coefficient, cm$^2$/kg | Max. deviation from linearity |
|---|---|---|
| Aluminum | $-3.8$ to $-4.2 \times 10^{-6}$ | $-0.001$ |
| Antimony | $+11.1 \times 10^{-6}$ | Nonlinear |
| Bismuth | $+21.4 \times 10^{-6}$ | Nonlinear |
| Cadmium | $-9.1 \times 10^{-6}$ | $-0.006$ |
| Copper | $-1.8 \times 10^{-6}$ | $-0.0004$ |
| Iron | $-2.3 \times 10^{-6}$ | $-0.0005$ |
| Lithium | $+7.72 \times 10^{-6}$ | |
| Manganin | $+2.3 \times 10^{-6}$ | |
| Mercury; | | |
| Liquid below 6,500 kg/cm$^2$ | $-22.4 \times 10^{-6}$ | |
| Solid above 7,600 kg/cm$^2$ | $-23.6 \times 10^{-6}$ | |
| Platinum | $-1.9 \times 10^{-6}$ | $-0.003$ |
| Rubidium | $-61 \times 10^{-6}$ | Nonlinear |
| 0–1,000 kg/cm$^2$ | $-155 \times 10^{-6}$ | Nonlinear |
| Silver | $-3.3 \times 10^{-6}$ | $-0.001$ |
| Sodium | $-37 \times 10^{-6}$ | Nonlinear |
| 0–1,000 kg/cm$^2$ | $-60 \times 10^{-6}$ | Nonlinear |
| Strontium, at 50°C | $58.3$ to $61.5 \times 10^{-6}$ | |

* P. W. Bridgman, *Proc. Natl. Acad. Sci. U.S.*, **3**, 10 (1917).

A pressure-sensing element for pressures up to 100,000 atm is described by Hall.[2] The calibration obtained at room temperature is expected to be reasonably valid under high-temperature operation. The change of resistance for most metals is of the order of 10 per cent for a change of pressure of 10,000 kg/cm$^2$; for manganin it is 2.3 per cent. The linear relationship of Eq. (1) is, of course, only an approximation. The maximum deviation from linearity in the pressure

[1] P. W. Bridgman, *Proc. Natl. Acad. Sci. U.S.*, **3**, 10 (1917).
[2] H. T. Hall, *Rev. Sci. Instr.*, **29**, 267 (1958).

range from 0 to 12,000 kg/cm² can be found in Table 6, column 3. The resistance variation follows the pressure variation without time lag or hysteresis. However, if the pressure is transmitted to the transducer by a liquid, a time lag may arise because of the increased viscosity of the liquid at high pressures.

A practical pressure transducer has been described by Bridgman.[1] It consists of a ring-shaped insulating core (bone) 1 cm in diameter and 0.5 cm thick; an insulated manganin wire (no. 38, approximately 30 ohms/m) is wound in a noninductive fashion around the core. The total transducer resistance is 100 ohms. The contact resistance should be small and should not change with pressure. Of course, the influence of the contact resistance is small if the resistance of the transducer itself is large.

The wire must be "seasoned" by successive pressure applications. Even after continued use, Bridgman observed a zero point shift (i.e., a change of resistance at 1 atm pressure) of $\frac{1}{2}$ per cent over the period of a month of continued usage. Periodic calibration is advised. Bridgman estimated an error of $\frac{1}{10}$ per cent in measuring pressure from 0 to 13,000 kg/cm² and in the temperature range from 0 to 50°C caused primarily by the reading error of the instruments.

The advantages of the resistive pressure transducer over any mechanical gauge are its simplicity and ruggedness. The introduction of insulated leads into the pressure chamber causes considerable technical difficulties at very high pressures.

### 1-52. CAPACITIVE PRESSURE TRANSDUCERS

The dielectric constants of gases, liquids, and solids vary with the pressure and can be measured by capacitive systems. A simple

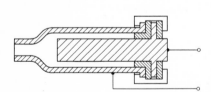

arrangement for capacitive pressure measurements in benzene has been described by Trendelnburg[2] and is shown in Fig. (1-5)2. It consists of a cylindrical capacitor suitable to withstand the pressure. In the range between atmospheric pressure and 100 atm, the dielectric constant changes by $\frac{1}{2}$ per cent. The static transfer function (change

FIG. (1-5)2. Capacitive pressure transducer based upon the variation with pressure of the dielectric constant.

of capacitance versus pressure) is nonlinear, as expected from theoretical considerations (Clausius-Mosotti equation). With the help of

[1] P. W. Bridgman, *Proc. Am. Acad. Arts Sci.*, **47**, 321 (1911).

[2] F. Trendelnburg, *Z. tech. Physik*, **11**, 465 (1930).

a compensating nonlinear network the author has succeeded in obtaining an output current that is a linear function of the pressure. The dynamic transfer function is lower by about 20 per cent than the static transfer function because of the variation of temperature which accompanies the adiabatic compression.

The system has been used in a pressure range from 0 to 200 kg/cm². Water and gases dissolved in the benzene may cause a hysteresis of the pressure-capacitance characteristic. The method is primarily applicable to nonpolar substances.

The capacitive method can be used to measure pressure in gases; the dielectric constant of air (19°C) changes from 1.0006 at 1 atm to 1.0548 at 100 atm.

Variations with pressure of the dielectric constants of solids (Rochelle salt) have been reported by D. Bancroft, *Phys. Rev.*, **53**, 587 (1938); of barium titanate by B. M. Sul and L. F. Vereshchagin, *Compt. rend. acad. sci. U.R.S.S.*, **48**, 634 (1945); of ionic crystals (MgO, LiF, NaCl, KCl, KBr) by S. Mayburg, *Phys. Rev.*, **79**, 375 (1950). Pressure cells for the determination of the dielectric constants at elevated pressure that can be used for pressure measurements are described by M. G. Vallauri and P. W. Fosbergh, Jr., *Rev. Sci. Instr.*, **28**, 198 (1957), and D. W. McCall, *Rev. Sci. Instr.*, **28**, 345 (1957).

1-53. PIEZOELECTRIC PRESSURE TRANSDUCER

Certain piezoelectric crystals, if exposed to pressure from all sides ("hydrostatic pressure"), will develop a polarization in a preferred crystal direction; the polarization charge gives rise to an output voltage. Piezoelectric pressure transducers are primarily useful for the measurement of transient pressures of the order of several thousand atmospheres and of a time range from a fraction of a second to more than ten microseconds.

Not all piezoelectric crystals are hydrostatically sensitive; primarily tartaric acid, tourmaline, and sucrose have been used for pressure transducers. The sensitivity ranges from $5.6 \times 10^{-12}$ coulomb per (lb/in.²) for sucrose to $21 \times 10^{-12}$ coulomb per (lb/in.²) for tartaric acid. The frequently used tourmaline has a sensitivity of $11 \times 10^{-12}$ coulomb per (lb/in.²). The output voltage is a linear function of the pressure.

A very simple piezoelectric pressure transducer consisting of a spark plug on which a crystal is mounted is described by Lawson and Miller.[1] Arons and Cole[2] have described the construction and investigation of a piezoelectric pressure transducer for the measurement of

[1] A. W. Lawson and P. H. Miller, *Rev. Sci. Instr.*, **13**, 427 (1942).
[2] A. B. Arons and R. H. Cole, *Rev. Sci. Instr.*, **21**, 31 (1950).

large transient pressures. The system is illustrated in Fig. (1-5)3 and consists of four circular tourmaline disks stacked on both sides of a steel plate. The tourmaline disks are coated with a conductive material and connected to the terminals. The entire assembly is coated with an insulating protective layer. Gauges of this type have been used for the measurement of pressure amplitudes ranging from 0.03 to 6,700 atm (0.5 to 100,000 lb/in.²).

Because of the small size of the pressure transducer (about $\frac{1}{2}$ in. diameter), the distortion of the pressure field is usually negligible. The distortion can be further reduced by orienting the disk-shaped transducer edgewise to the propagation direction of the pressure wave.

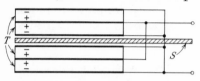

FIG.   (1-5)3. Piezoelectrical pressure transducer [*from A. B. Arons and R. H. Cole, Rev. Sci. Instr.*, **21**, 31 (1950); *by permission*].

The output impedance is high. An equivalent circuit for a piezoelectric transducer is shown in Fig. (1-2)64. The dynamic response of the transducer falls off at low frequencies and becomes zero for static pressures for reasons explained in 1-25. The response to high-frequency pressure fluctuation can extend over a considerable range. In the system described by Arons and Cole, the response begins to fall off between 10 and 20 kc and reaches zero in the vicinity of 100 kc.

Change of the crystal temperature causes spurious output voltages (pyroelectric effect). The error can be considerable; in the system described by Arons and Cole, a change in temperature of 1°C produces a charge equivalent to that caused by a pressure change of 13 atm. However, a temperature change of this order of magnitude occurs rarely during the short time of a pressure-transient measurement and can also be kept low by thermal insulation of the transducer. Undesirable signals are also likely to arise in the cables connecting the pressure-sensing element with the accessory equipment. The effect has been studied by Cole.[1]

## 1-54. ELECTROKINETIC PRESSURE TRANSDUCER

If a liquid such as water is brought in contact with a solid body such as a glass particle, an electric field of molecular thickness will arise at the interface between the two substances. Under the influence of this field, the glass as well as some of the water molecules

---

[1] R. H. Cole, *Rev. Sci. Instr.*, **21**, 32, footnote 3 (1950).

will acquire a net charge, the glass being negatively, the water molecules positively, charged. A movement of the glass particles through the water constitutes a current (streaming current) which can be measured. Conversely, a movement of the liquid through a stationary fritted glass disk gives rise to a potential difference (streaming potential) between both sides of the fritted disk.

This phenomenon forms the basis of the electrokinetic pressure meter. The system is illustrated in Fig. (1-5)4 and consists of a porous porcelain disk and a container filled with a polar liquid (aceto-nitrile) on either side. If the liquid is pressed through the disk, a potential difference arises between both sides of the porcelain disk which can be picked up by means of two wire-mesh electrodes.

To measure static pressure, the liquid would have to be pressed continuously through the disk. If alternating pressure is applied, the liquid will be pressed back and forth, and an alternating voltage will arise between the electrodes. A commercial form of this pressure transducer operates over a range from $10^{-4}$ to $10^2$ psi. The output is a linear function of the applied

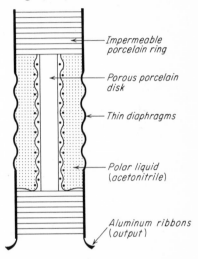

Fig.  (1-5)4. Electrokinetic pressure transducer [*from C.E.C. Recording*, **10**, 6 (1956), *Consolidated Electrodynamics Corp., Pasadena, Calif.; by permission*].

pressure difference. The sensitivity is 350 mV/psi under no-load condition and about 3 $\mu$A/psi under short-circuit condition. The output impedance is about $10^5$ ohms, shunted by a capacitance of 40 $\mu\mu$F $\pm$ 10 $\mu\mu$F, but can be made to have values up to several hundred megohms.

The output follows the applied pressure without delay. The dynamic response characteristic is flat within 1 dB, from 4 to 15,000 cps and within 2 dB from 3 to 25,000 cps. The output varies slightly with temperature; the output voltage changes by about 6 per cent if the temperature is changed from 70 to $-10°$F or to $+140°$F.

For references see M. Williams, *Rev. Sci. Instr.*, **19**, 640 (1948); also *C.E.C. Recordings*, **10**(6), 6 (1956), and *Bulletin* 1573, both of the Consolidated Electrodynamics Corporation, Pasadena, Calif.

### 1-55. PRESSURE TRANSDUCERS BASED UPON ELECTRIC GAS DISCHARGES

The most frequently reported electric gas discharge form for air-pressure and flow-velocity measurements is the corona discharge. The method is described by Werner[1] and Werner and Geronime.[2] The system is illustrated schematically in Fig. (1-5)5 and consists of a fine needle point electrode $A$ (Wollaston wire) and a large electrode

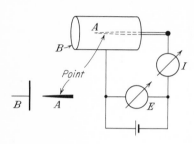

$B$ which may either surround the electrode $A$ or may be a flat or a cylindrical electrode located in front of the needle point and separated by several millimeters from the point. If a voltage $E$ of the order of 1,000 to 5,000 volts is applied, a corona discharge will form in the vicinity of the needle point, and a current $I$ up to ten or more microamperes will be observed. The general form of the voltage-current characteristics for positive and negative points is shown in Fig. (1-5)6.

FIG. (1-5)5. Pressure transducer based upon electric gas discharges; schematic diagram of point-to-plane and concentric electrode configurations.

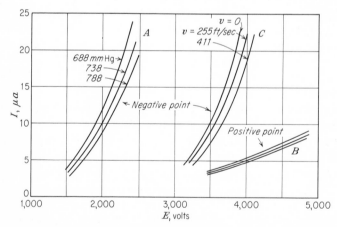

FIG. (1-5)6. Corona discharge characteristics. Curves $A$ and $B$ for point-to-plane corona, 3 mm spacing, at different pressures; curve $C$ (other geometry), at different velocities [*from F. D. Werner, Rev. Sci. Instr.*, **26**, 61 (1950); *by permission*].

[1] F. D. Werner, *Rev. Sci. Instr.*, **21**, 61 (1950).

[2] F. D. Werner and R. L. Geronime, *Tech. Rept.* 53-142, Wright Air Development Center, Wright-Patterson Air Force Base, Ohio, 1953.

The voltage-current characteristic is different for positive or negative polarity of the point and changes strongly with the pressure, and, to a lesser extent, with the gas velocity. The influence of humidity is slight; that of temperature negligible. The instrument, therefore, measures primarily the pressure variation in the flowing gas. The characteristic is also influenced by the geometry of the needle point.

The response of the system is extremely fast; pressure variations in the megacycle region have been observed. In the range from $10^2$ to $10^6$ cps, the relative response does not change; the frequency-response curve is flat throughout this range. The noise level can be considerable.

Other discharge forms, such as the Townsend discharge or the glow discharge, have also been investigated as a means to measure velocity and pressure.

For a summarizing review, see W. M. Cady, Physical Measurements in Gas Dynamics and Combustion, in R. W. Ladenburg (ed.), "High Speed Aerodynamics and Jet Propulsion," vol. 9, sec. C, 3, p. 146, Princeton University Press, Princeton, N.J., 1954. For references see Werner, *loc. cit.*, as well as Werner and Geronime, *loc. cit.*

### 1-56. THERMAL GAUGES

The following thermal transducer is designed primarily for operation below atmospheric pressure (i.e., as a vacuum gauge). It is illustrated in Fig. (1-5)7 and it usually contains a fine wire $F$ of about 0.001 in. diameter mounted in a glass or metal tube. The tube is joined to the system where the pressure is to be measured. The wire is heated to a temperature between 75 and 400°C by a current of the order of 10 to 100 mA and cooled by heat conduction through the gas surrounding the wire.

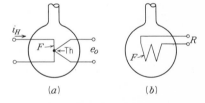

(a)    (b)

Fig. (1-5)7. Thermal vacuum gauges: (a) the temperature of the heated filament $F$ is measured by the thermoelement $Th$; (b) the temperature of the filament is measured by its resistance variation.

At atmospheric pressures the heat conductivity of gases is independent of their pressures. However, as the pressure is reduced, a point is reached (about 10 mm Hg) at which the heat conductivity begins to decrease with decreasing pressure. The gas molecules striking the hot wire carry off energy to the walls of the vessel; the actual temperature of the wire is then controlled by the rate at which gas molecules collide with it. As the pressure is lowered, the wire will lose heat less rapidly because there are fewer molecules available for heat transfer; consequently its temperature will increase.

Cooling of the wire is the result not only of thermal conduction through the gas; radiation as well as conduction through the supports of the wire also play a part which increases in relative significance as the pressure is lowered. At less than $10^{-3}$ mm the loss by radiation and conduction through the supports begins to get larger than the heat loss by conduction through the gas.

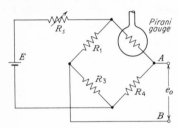

FIG. (1-5)8. Circuit diagram of a Pirani gauge.

The temperature of the filament can be measured in two ways: Either a thermoelement is hooked on the filament and furnishes an output proportional to the filament temperature, as in the thermo-couple gauge, Fig. (1-5)7a, or the filament is made of a material with a high resistance-temperature coefficient (tungsten, platinum, nickel) and the variation of resistance caused by a variation of the filament temperature is measured, as in the Pirani gauge, Fig. (1-5)7b.[1]

The Pirani gauge is commonly connected in a resistance bridge as shown in Fig. (1-5)8. The voltage or the current applied to the bridge is kept constant and at a given pressure in the gauge the bridge is balanced by adjusting $R_1$, $R_3$, or $R_4$. A variation of the pressure causes a bridge unbalance, i.e., the appearance of a voltage between $A$ and $B$.

Another way of operating the gauge is the following: The bridge is continuously kept in balance. If the filament temperature changes in response to pressure variation in the gauge, such a change is compensated for by varying the current through the filament (adjustment of the resistance $R_s$). The change of current or voltage applied to the bridge or to the filament is noted and is a measure for the gas pressure surrounding the filament.[2] A feedback-controlled, self-balancing thermal-gauge system is described by Leck and Martin.[3] With this method the temperature of the gauge can be kept quite low (less than 75°C).

[1] M. Pirani and R. Neumann, *Electronic Eng.*, **17**, 227 (December, 1944); *ibid.*, 322 (January, 1945); *ibid.*, 367 (February, 1945); *ibid.*, 422 (March, 1945).

[2] N. R. Campbell, *Proc. Phys. Soc.* (*London*), **33**, 287 (1921).

[3] J. H. Leck and C. S. Martin, *Rev. Sci. Instr.*, **28**, 119 (1957).

Empirical calibration is required. Typical calibration curves are shown in Fig. (1-5)9. Since different gases have different heat conductivities, the calibration curve for each will be different. For some applications, notably in the absence of hydrogen, this difference may be neglected. The Pirani gauge indicates the total pressure, i.e., the sum of all partial pressures of any gases and vapors present.

The ordinary range of the gauge is from about 1 to $10^{-3}$ mm. Different investigators have extended the range to as high as 15 mm[1] and to as low as $10^{-5}$ mm Hg.[2] Extension to a limit of $5 \times 10^{-9}$ mm

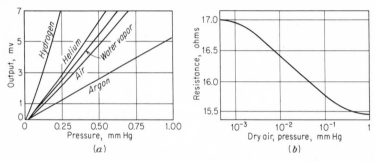

Fig. (1-5)9. Characteristics of thermal vacuum gauges: (*a*) output from a commercial thermocouple gauge for different gases; (*b*) resistance variation of a Pirani gauge.

has been claimed by Ellet and Zabel,[3] who used a flattened-nickel filament and a liquid-air bath surrounding the gauge, and by Weise,[4] who used a film of magnesium titanate 20 $\mu$ thick, with a surface of about 10 cm². This gauge has a range from 760 mm Hg to the lowest measurable pressure. At high pressures (above 200 mm Hg) the effect seems to reside in a cooling of the film brought about by a gas convection current created within the vessel.

Operation of the gauge over the full sensitive range causes, in general, a resistance variation of 10 per cent, e.g., from 15.5 to 17 ohms. The time required for a thermal gauge to come to equilibrium varies from several seconds to minutes.

Pirani gauges show a variation of reading with ambient temperature. In order to compensate for ambient-temperature variation of the gauge, a "dummy gauge," i.e., a gauge with similar physical properties but operated in a closed and evacuated vessel, should be connected in an adjacent bridge arm. A constant-temperature bath

[1] E. S. Rittner, *Rev. Sci. Instr.*, **17**, 113 (1946).
[2] A. M. Skellett, *J. Opt. Soc. Am.*, **15**, 56 (1927).
[3] A. Ellet and R. M. Zabel, *Phys. Rev.*, **37**, 1102 (1931).
[4] E. Weise, *Z. tech. Physik*, **24**, 66 (1947).

surrounding the gauge is sometimes recommended. Difficulties may arise if the gauge is used in an atmosphere of gases that may react with the hot filament.

The construction of very small Pirani gauges (less than 0.5 cc) is described by D. G. H. Marsden, *Rev. Sci. Instr.*, **26**, 1205 (1955). G. A. Slack, *Rev. Sci. Instr.*, **27**, 241 (1956), describes a gauge of unusual construction for the measurement of helium pressure at low temperature. For further modification and references see H. Schwarz, *Arch. tech. Messen*, V 1341–3, January, 1952.

### 1-57. IONIZATION GAUGES

*a. Thermionic Gauge.* The gauge system is constructed like a triode, i.e., it contains, in a glass vessel, a heated electron-emitting

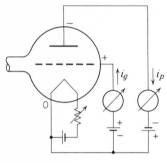

cathode, a grid surrounding the cathode, and a plate surrounding the grid. The grid is maintained at a positive potential (100 to 250 volts) with respect to the cathode, the plate at a negative potential (2 to 50 volts), as shown in Fig. (1-5)10.

Electrons emitted from the cathode are accelerated by the grid; they will collide with the gas molecules in the vessel and will ionize them. The positive ions in the space between grid and plate will be collected by the negative plate; those formed in the space between cathode and grid will migrate toward the cathode. The electrons and negative ions will be collected by the positive grid.

FIG. (1-5)10. Thermionic ionization gauge.

The rate of ion production is proportional to the amount of gas present and to the number of electrons available to ionize the gas. The ratio of positive ions, i.e., plate current $i_p$, to the grid current $i_g$ is, therefore, a measure of the pressure $p$ within the tube; i.e.,

$$p = k \frac{i_p^+}{i_g^-}$$

where $k$ is a constant of the order of magnitude of 10 $\mu$ Hg (mA/$\mu$A) varying with the geometry of the tube, nature of the gas, and operating voltages.[1] The ionization gauge measures the total pressure of all gases present. The indication is continuous and inertia-free.

Thermionic ionization gauges can be used for pressure measurements in the range between $10^{-3}$ and $10^{-8}$ mm Hg. Within this range the output varies almost linearly from $10^{-4}$ to $10^{-9}$ amp. If the

[1] S. Dushman, *Phys. Rev.*, **17**, 7 (1921), and **23**, 734 (1924).

pressure rises above $10^{-3}$ mm Hg, the impact of positive ions upon the cathode tends to heat and destroy the cathode.

At a pressure lower than $10^{-8}$ mm Hg, the following effect imposes a limit to the application: The impact of electrons upon the grid causes a soft X-ray radiation; these X rays liberate electrons from the plate.[1] At a pressure of less than $10^{-8}$ mm Hg the electron-emission current from the plate is of the same order of magnitude and indistinguishable from the ion current to the plate. By using a very small plate (a wire centrally suspended in the helical grid space; the filament is outside the grid helix), Bayard and Alpert[2] have succeeded in reducing the photoelectron current and in increasing the useful range of the ionization gauge to $10^{-10}$ mm Hg.[3] A modification and improvement of the Bayard-Alpert gauge is described

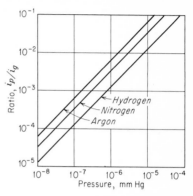

FIG. (1-5)11. Characteristics of a thermionic ionization gauge for different gases.

by Nottingham.[4] Another thermionic ionization gauge which permits measurements of pressure down to the order of $10^{-10}$ mm Hg is described by Lander.[5] Two gauge types for measuring pressures up to the millimeter range are described by Schulz and Phelps.[6]

Since the efficiency of the ionization process varies for different gases, the calibration will depend upon the type of gas in the gauge; impurities in the gas will introduce errors in the pressure measurement. Calibration curves for three different gases are shown in Fig. (1-5)11.

The usual method of operating the gauge consists in keeping the grid current constant (about 1 to 20 mA) and measuring the plate current. Various methods for keeping the grid current constant have been suggested. Hariharan and Bhalla[7] have described a system in which the ratio $i_p/i_g$ is measured with a logarithmic differential amplifier. This system, which forms an output current proportional

[1] W. B. Nottingham, *J. Appl. Physics*, **8**, 762 (1937).

[2] R. T. Bayard and D. Alpert, *Rev. Sci. Instr.*, **21**, 571 (1950).

[3] See also G. H. Metson, *Brit. J. Appl. Phys.*, **2**, 46 (1951).

[4] W. B. Nottingham, *Trans. Vacuum Symposium*, 1954, p. 76.

[5] J. J. Lander, *Rev. Sci. Instr.*, **21**, 672 (1950).

[6] G. J. Schulz and A. V. Phelps, *Rev. Sci. Instr.*, **28**, 1051 (1957).

[7] P. Hariharan and M. S. Bhalla, *Rev. Sci. Instr.*, **27**, 448 (1956).

to $\log i_p^+ - \log i_g^-$, eliminates the need for accurately stabilizing the electron current and permits the reading of the pressure in the range from $10^{-3}$ to $10^{-7}$ mm Hg on a single logarithmic scale. The maximum deviation from a true logarithmic response is 3 per cent of the full-scale value.

The influence of the filament, grid, and plate geometry upon the sensitivity of conventional ionization gauges has been investigated by Kinsella.[1] A relative optimum has been found for a grid structure of 1 cm diameter with a pitch of 4 mm, 0.2 to 0.3 mm wire diameter; the sensitivity of the gauge increases with the plate diameter.

Application of excessively high grid or plate voltages to the gauge may lead to instability (Barkhausen-Kurz oscillations).

Difficulties are likely to arise if gases within the gauge react with the cathode or with other elements of the gauge (e.g., water vapor, carbon monoxide, carbon dioxide, halogen gases, and hydrocarbon vapors). Oxidation of the filament or gas absorption may occur if the gauge is operated at a pressure higher than $10^{-3}$ mm Hg. Under normal conditions a filament life of $10^3$ to $10^4$ hr can be expected.

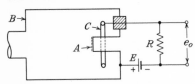

Fig. (1-5)12. Radioactive ionization gauge. *A*, α source; *B*, housing; *C*, ion collector; *E*, voltage source (*alphatron, model 511, National Research Corp., Cambridge, Mass.; by permission*).

Outgassing of the gauge, in particular of the grid by passing a high current through it, is required when the system has been subjected to a pressure exceeding $10^{-2}$ mm Hg. Some gauges are constructed with a grid and a plate in the form of helical filaments which can be heated and outgassed by passing a current through each helix. Extended overloading during outgassing may cause evaporation of metal on the glass wall and may lead to leakage paths.

For bibliography, see H. Schwarz, *Arch. tech. Messen*, V 1341–5, May, 1952; also *ibid.*, V 1341–2, September, 1951, and V 1341–4, March, 1952.

*b. Radioactive Ionization Gauge.* The instrument is shown schematically in Fig. (1-5)12. It contains a radioactive source *A* that emits α particles which ionize the gas in the gauge. The number of ions formed in the gas is directly proportional to the gas pressure as long as the range of the α particles is longer than the dimensions of the chamber. The ions are collected at an electrode *C* and form a current of the order of $10^{-9}$ to $10^{-13}$ amp. In the commercial model (Alphatron, Model 511, National Research Corporation, Cambridge,

[1] J. J. Kinsella, *Trans. Vacuum Symposium*, 1954, p. 65.

Mass.) a vacuum-tube electrometer is used to detect the current by measuring the voltage drop across the resistance $R$.

The transfer characteristic (meter reading versus pressure) is linear. The error, i.e., the practical deviation from the characteristic, is 2 per cent. The slope of the characteristics is different for different gases, as shown in Fig. (1-5)13. This has the advantage that the gauge can be used as a leak detector. The gauge will show an indication if a gas with a relative response different from air diffuses through a leak into the gauge (e.g., $H_2$ or He). On the other hand, the different calibration characteristics for different gases can cause uncertainties and errors when impurities of an unknown concentration or composition are present.

The system can be used over a range from $10^3$ to $10^{-3}$ mm Hg ($10^{-3}$ mm Hg is 10 per cent of lowest full-scale reading). At lower gas pressures the mean free path of the $\alpha$ particles increases beyond the dimensions of the vessel, so that the probability of a collision is reduced and the ionization current is very small. X-ray effects as described under Thermionic Ion Gauges, 1-57a, ultimately limit the useful range of the gauge.

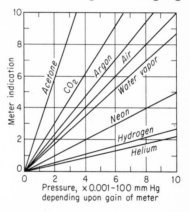

FIG. (1-5)13. Characteristics of a radioactive ionization gauge (*alphatron, model* 511, *National Research Corp., Cambridge, Mass.; by permission*).

An advantage of the Alphatron over the thermionic ionization gauge is the absence of the heated filament. The instrument cannot be damaged by exposure to air of atmospheric pressure and requires no degassing. The indication is continuous and follows the variation of the pressure without inertia. The activity of the $\alpha$ source (radium-gold alloy, rhodium-plated to prevent loss of the short-lived gaseous daughter products of radium, radon) will rise slightly as the source comes to equilibrium with its daughter products over the course of about fifty years. For practical purposes the radiation from the source will be constant within a few per cent for a period of five years.

For references see J. R. Downing and G. Mellen, *Rev. Sci. Instr.*, **17**, 218 (1946); also *Electronics*, **19**, 142 (April, 1946); G. Mellen, *Ind. Eng. Chem.*, **40**, 787 (1948); A. Boburieth, *Le Vide*, **1**, 61 (1946); J. H. Beynon and G. R. Nicholson, *J. Sci. Instr.*, **33**, 376 (1956). A similar instrument in which ionization by beta rays takes place is described by C. Deal et al., *Anal. Chem.*, **28**, 1958 (1956).

*c. Philips-Penning Ionization Gauge.* This gauge is shown in Fig. (1-5)14. A high voltage is applied to the electrodes $C_1$, $C_2$, and $A$ and causes a discharge in the gas. The discharge current depends upon the electrode geometry, the type of gas in the gauge, and the gas pressure, and can be used as a means for the pressure determination.

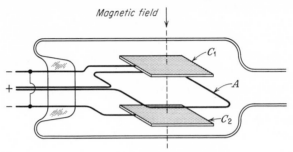

Fig. (1-5)14. Philips-Penning gauge.

At a pressure below $10^{-2}$ mm Hg, the mean free path becomes so large that the probability of a collision and ionization is too small to maintain a discharge. By applying a magnetic field and by using the geometric arrangement shown in Fig. (1-5)14, Penning[1] has succeeded in increasing the path length of the electrons and in maintaining a discharge beyond $10^{-5}$ mm Hg.

Calibration curves of a Philips-Penning gauge for different gases are shown in Fig. (1-5)15. The useful range of the gauge extends from $10^{-3}$ to $10^{-5}$ mm Hg. Several authors have succeeded in extending this range. Penning and Nienhuis[2] have technically improved the gauge and been able to use it down to $4 \times 10^{-7}$ mm Hg. Hayashi[3] has extended the range to lower pressures ($10^{-6}$) and, in another modification by reducing the size of the anode ring, to higher pressure ($10^{-1}$ mm Hg). However, at these high pressures a considerable amount of sputtering occurs which leads to electric

Fig. (1-5)15. Characteristics of a Philips-Penning ionization gauge for various gases.

[1] F. M. Penning, *Physica*, **4**, 71 (1937), and *Philips Tech. Rev.*, **2**, 201 (1937).

[2] F. M. Penning and K. Nienhuis, *Philips Tech. Rev.*, **11**, 116 (1949).

[3] C. Hayashi et al., *Rev. Sci. Instr.*, **20**, 524 (1949).

leakage problems. Readhead[1] has succeeded in developing a gauge (inverted magnetron) which is capable of measuring pressure in a range from $10^{-3}$ to $10^{-12}$ mm Hg.

The gauge is generally operated with direct current of 1,000 to 3,000 volts. It can be driven with alternating current since the different geometry of the electrodes causes rectification, but the a-c discharge extinguishes at a pressure below $2 \times 10^{-5}$ mm Hg. Starting difficulties are encountered, even at d-c operation if the gauge is turned on at low pressures. According to McIlwraith,[2] such difficulties can be overcome by momentarily flashing a filament that is sealed in a side tube close to the gauge while the high voltage is turned on.

The magnetic fields used by different authors vary from 300 to 8,000 oersteds. The accuracy of a calibrated gauge is in general of the order of 3 per cent.

## 1-6. Transducers Responding to Flow Velocity of Liquids

The only direct-acting flow-velocity transducer, in the strict sense of the word "direct," is the inductive system described in 1-62. All others can be considered as indirect (mechanoelectric) systems, such as the resistive systems (1-61), the sonic system (1-64), or the radioactive methods (1-65). The thermoelectric system (1-63) is also, in principle, an indirect-acting system.

A considerable number of methods for flow-velocity measurements have been described which are combinations of mechanical systems with electrical-displacement transducers. For instance, the flowing medium can be made to cause a pressure difference in a pitot tube or a restricted orifice and the pressure is measured with electric transducers, or the flowing medium causes a displacement of the float in a rotameter which is transformed in an electric signal.[3] Other methods use rotating impellers driven by the liquid medium; either the torque caused by the flow or the rotation velocity of the impeller is picked up with electric transducers. These combination systems are omitted from the description. A survey (mostly mechanical) of basic solutions for flow measurement can be found in Ziebolz.[4]

Linearity between the flow velocity and the transducer output

---

[1] P. A. Readhead, paper read at the Physical Electronics Conference, M.I.T., March, 1957.

[2] C. G. McIlwraith, *Rev. Sci. Instr.*, **18**, 683 (1947).

[3] E. C. Crittendon and R. E. Shipley, *Rev. Sci. Instr.*, **15**, 343 (1944).

[4] H. Ziebolz, *Rev. Sci. Instr.*, **15**, 80 (1944).

signal is of particular importance when the time average of an intermittent or pulsating flow velocity is to be measured. If the flow
rate is measured by means of a nonlinear system, such as an orifice
and a transducer which measures the pressure difference $p$ before
and in the orifice, and if the flow rate varies with time, the average
flow rate $F$ can be expressed by

$$F = c(\sqrt{p})_{\text{av}}$$

Since there is always an inertia in the measurement of $p$, it is likely
that not the instantaneous pressure $p$ but the average pressure $p_{\text{av}}$ is
measured, so that

$$F' = c\sqrt{p_{\text{av}}}$$

$F'$ is always larger than $F$, sometimes by as much as 40 per cent.

Some of the methods for measuring liquid flow can be used for the
flow-velocity measurement of gases as well, e.g., the thermal-, the
sonic-, and the radioactive-tracer methods. Anrep and Downing[1]
have used a *gas* flow-velocity transducer such as described in 1-66 for
*liquid* flow-velocity measurements. The liquid enters a flask and
displaces the air from the flask. The velocity of the escaping air is
measured with a thermal transducer.

### 1-61. Resistive Transducer Systems for Liquid Flow Velocity

*a. Electrolytic-tracer Method.* This method, which is suitable only
for discontinuous measurements of the linear flow velocity and the
volume flow rate of water, has been described by Allen and Taylor[2]
and is illustrated in Fig. (1-6)1. The water flows through a conduit
with the average linear velocity $v$. At the point $N$, a shot of a salt
solution is injected into the stream, causing a cloud of electrolyte with
a conductivity higher than that of the surrounding water. The cloud
moves with the water; the passage of the conductive cloud causes an
increase of an electric current through the electrodes $AA$ and, later,
the electrodes $BB$, which are spaced from the electrodes $AA$ by a
distance $L$. The time difference $t$ between these passages is measured
and the velocity is determined from $\bar{v} = L/t$. If $V$ is the volume of
the water between the two pairs of electrodes, the flow rate $F$ can be
determined from $F = V/t$. The method can be modified by measuring the time difference between the injection of the solution and the
passage of the solution cloud through one pair of electrodes.

[1] G. V. Anrep and A. C. Downing, *J. Sci. Instr.*, **3**, 221 (1925).
[2] C. M. Allen and E. A. Taylor, *Trans. ASME*, **45**, 285 (1923).

The method has been primarily applied for large conduits (order of magnitude of $L$ is 50 to several hundred meters) and for flow rates of the order of several cubic meters per second. The salt-solution injector system for such large channels can be quite complex.

Errors are caused through differences in velocity of the particles in the conductive cloud (longitudinal growth of the salt cloud as it moves downstream), leading to uncertainties in the determination of the time difference $t$, since the passage of the cloud through the electrodes $AA$ or $BB$ is not instantaneous. Allen and Taylor recommend the elimination of this error by a graphic method in which the time difference $t$ is measured between the center of gravity of the current-time areas as indicated in Fig. (1-6)1. Further errors may arise through inhomogeneous distribution of the particle velocity and variation of concentration over the cross section. De Haller in a theoretical study[1] indicates that this error diminishes with increased turbulence (increased Reynolds num-

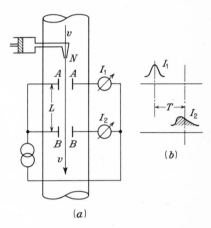

Fig. (1-6)1. Flow-velocity transducer for liquids: (*a*) electrolytic tracer system; (*b*) current-time function at the electrodes $AA$ and $BB$.

ber). The error for laminar flow is considerably larger than for turbulent flow. In order to reduce the error resulting from the variation of velocity over the cross section, de Haller proposes the use of a point electrode positioned at a distance of 0.75 to $0.8R$ ($R =$ internal radius of the conduit) from the center of the conduit if the velocity profile is axially symmetric, or of a ring electrode of $0.77R$ if the velocity profile is asymmetric. Other authors consider screens covering the entire cross section of the conduit as the best form of electrodes.

Experimental studies of the salt-velocity method have been carried out by Müller.[2] In a comparative study on flow meters, Kirschmer and Esterer[3] found that the total error of the method is between 1.2 and 6 per cent. The error is always negative, i.e., the actual flow rate is larger than that indicated by the method. The

[1] P. de Haller, *Helv. Physica Acta*, **3**, 17 (1930).

[2] K. E. Müller, *Schweiz. Bauz.*, **87**, 41 (1926).

[3] O. Kirschmer and B. Esterer, *Z. Ver. deut. Ing.*, **74**, 1499 (1930).

method of injecting the salt solution (single and multiple injectors, symmetric and asymmetric) is of negligible influence.

The electrolytic-tracer method has been used by Prausnitz and Wilhelm[1] for measuring the turbulent concentration fluctuations and, thus, the variation with time of turbulence in a liquid. An electrolyte (hydrochloric acid-methanol mixture) is continuously injected into a stream of tap water. The conductivity fluctuations are measured downstream by a very small probe consisting of two ends of platinum wire (1.2 mm long and 1 mm apart) protruding from the end of a tapered glass tube. The output signal (variation of resistance with time) is electronically processed and furnishes information for the study of turbulence.

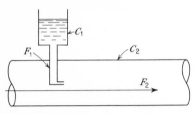

FIG. (1-6)2. Flow-velocity transducer for liquids, electrolytic dilution method.

*b. Electrolytic-dilution Method.*

A salt solution of a concentration $C_1$ is introduced at a steady and known flow rate $F_1$ into a stream of a liquid of low conductivity (water) flowing with the flow rate $F_2$. After mixing, the concentration is

$$C_2 = C_1 \frac{F_1}{F_2} \qquad (1)$$

The concentration $C_2$ is determined at a point downstream by electrolytic-resistance measurement (or by any other chemical or optical method), and the unknown flow rate $F_2$ is determined from Eq. (1).

The method has been investigated by Kirschmer and Esterer, *op. cit.*, p. 1505, and found to be very accurate; the error for large flow rates was found to vary between 0.03 and 0.06 per cent.

### 1-62. LIQUID-FLOW INDUCTION TRANSDUCER

The liquid, which should have a conductivity of at least $10^{-5}$ ohm$^{-1}$-cm$^{-1}$, flows through an insulating tube as shown in Fig. (1-6)3. The tube lies in a magnetic field of a flux density $B$. Two electrodes are inserted into the tube, their surfaces flush with the inner surface of the tube. If the liquid flows through the tube with the average velocity $\bar{v}$ (averaged over the cross section of the tube), charges within the liquid are moved to the electrodes, and a potential difference arises between both electrodes which is

$$e_o = Bd\bar{v} \times 10^{-8} \qquad \text{volts} \qquad (1)$$

[1] J. M. Prausnitz and R. H. Wilhelm, *Rev. Sci. Instr.*, **27**, 941 (1956).

where $B$ is the flux density in gauss, $d$ the distance in centimeters between the electrodes, i.e., the inner diameter of the tube, and $\bar{v}$ the average linear velocity of the liquid in centimeters per second. Equation (1) can be transformed into

$$e_o = \frac{BF}{d} \frac{4}{\pi} \times 10^{-8} \qquad \text{volts} \qquad (2)$$

where $F$ is the average flow rate of the liquid in milliliters per second.

The method has been perfected primarily by the work of A. Kolin and collaborators. See literature references until 1951 in W. G. James, *Rev. Sci. Instr.*, **22**, 989 (1951).

The magnetic field is usually of the order of 1,000 oersteds and may be continuous (direct current) or alternating (alternating current). In the first case the output voltage $e_o$ will be a d-c voltage; in the latter case it will be an a-c voltage, and $e_o$ will denote the instantaneous value if $B$ denotes the instantaneous value of the flux density. The a-c method reduces the danger of electrolytic polarization at the electrodes if the frequency is sufficiently high and permits the use of a-c amplifiers in subsequent stages. At the same time, it introduces in the output parasitic a-c voltages from capacitive a-c

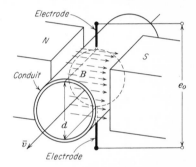

Fig. (1-6)3. Flow-velocity transducer based upon electromagnetic induction in liquids.

pickup and from asymmetric induction in the liquid and in the leads. The magnitude of these voltages increases with the frequency.

The transducer can be used for flow rates from the order of $10^{-3}$ mm/sec up to the highest values. According to Thürlemann[1] and Kolin,[2] the transducer output is independent of the type of flow (laminar or turbulent) and of the form of the velocity profile. However, Shercliff[3] has shown that any axial asymmetry may severely influence the output. If the flow velocity near one electrode is appreciably higher than the average velocity, the sensitivity of the transducer ($S = \Delta e_o / \Delta \bar{v}$) can be increased by a factor as high as 2, and if the flow is concentrated near the wall between both electrodes, the

[1] B. Thürlemann, *Helv. Physica Acta*, **14**, 383 (1942).
[2] A. Kolin, *Rev. Sci. Instr.*, **16**, 109 (1945).
[3] J. A. Shercliff, *J. Appl. Physics*, **25**, 817 (1954), and *J. Sci. Instr.*, **32**, 441 (1955).

sensitivity can be reduced by a factor of less than 0.5 or can even be negative. Even at axially symmetric flow, the sensitivity is 7.4 per cent lower than that computed from Eqs. (1) and (2). In narrow rectangular channels the sensitivity is (according to Shercliff) independent of the velocity profile. In such conduits the electrodes should extend over the entire width of the channel.

The output voltage is in general in the microvolt range. A particular sensitive and accurate instrument is described by James;[1] it has an input range from zero to about 40 mm/min and furnishes, at the transducer output, an a-c voltage (11.4 cps) of 0.5 $\mu$V/(ml)(min) $\pm$ 1 per cent for liquids of an electric conductivity of at least $3.5 \times 10^{-6}$ ohm$^{-1}$-cm$^{-1}$. The electric conductivity of the liquid determines the output impedance which may vary from several ohms, for mercury, up to several megohms.

The output is proportional to the flow velocity; the deviation from linearity can be kept small and corresponds, in the above-mentioned instrument of W. G. James, to about 0.25 ml/min. The output follows the variations of the flow velocity without inertia. Flow transients of 1 msec duration have been measured by Arnold.[2]

Errors may arise from parasitic voltages between the electrodes. Such voltages are independent of the flow and may stem from galvanic potentials between the electrodes and other metallic parts in the system[3] or from polarization of the electrodes at d-c operation. Parasitic d-c voltages due to flow (with the magnetic field removed) have also been observed. They can be eliminated by the use of graphite electrodes. The use of nonpolarizable electrodes is generally required at d-c operation. The parasitic a-c voltages mentioned above can be canceled by an auxiliary compensating voltage of appropriate magnitude and phase. The amount of random noise originating in the transducer and the influence of stray electric and magnetic fields increases, of course, with increased resistivity of the liquid.

The advantages of the method are reliability, simplicity, and ruggedness. There are no moving parts or constrictions in the tube. The response is fast, it is linear, and the output is independent of the physical properties of the liquid, with the exception of the electrical conductivity. Disadvantages of the method are the small output voltage, which requires amplification, and in some cases the high

[1] W. G. James, *Rev. Sci. Instr.*, **22**, 989 (1951).

[2] F. S. Arnold, *Rev. Sci. Instr.*, **22**, 43 (1951).

[3] A. J. Morris and J. H. Chadwick, *Trans. AIEE, Conf. Paper* T–1–58, 346 (1951).

output impedance, which increases the influence of external disturbances. The requirements for accessory equipment can be considerable. A commercial inductive flow meter is shown in Fig. (1-6)4.

### 1-63. THERMAL FLOW-VELOCITY TRANSDUCER

The method devised for the study of blood flow velocity in animal experiments ("Thermo-Stromuhr") by Rein[1] is analogous to that

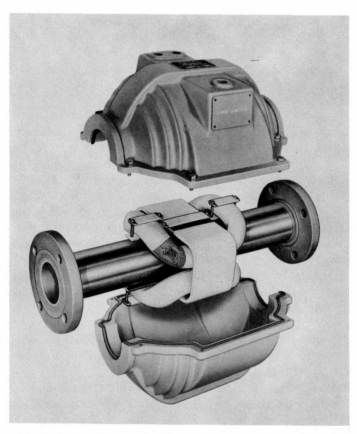

FIG. (1-6)4. Exploded view of an electromagnetic-induction transducer for flow-velocity measurement in liquids (*courtesy of The Foxboro Company, Foxboro, Mass.*).

described earlier (1911) by C. C. Thomas for the measurement of flow velocity in gases (1-66).

The instrument is illustrated schematically in Fig. (1-6)5. The blood flowing in a vessel $A$ is heated in an rf electric field between the

[1] H. Rein, *Z. Biol.*, **87**, 394 (1928).

two capacitor plates $C_1$ and $C_2$. The rf power—and thus the rate of heat generated in the blood—is kept constant. The longer the blood remains in the electric field between the capacitor plates, i.e., the lower its flow velocity, the more it will heat up. The increase in blood temperature, i.e., the temperature difference $\Delta t$ between a point upstream and one downstream from the heating device, is measured by two thermoelements $T_1$ and $T_2$ in thermal contact with the outside of the vessel. If $P$ is the electric power converted into heat, the flow velocity $F$ of the blood can be found from

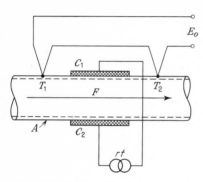

FIG. (1-6)5. Flow-velocity transducer for liquids, thermal method.

$$F = k\,\frac{P}{\Delta t}$$

where $k$ is an apparatus constant.

The system measures the average mass rate of flow. The transfer function (thermoelement output $E_o$ versus flow rate $F$) is approximately hyperbolic.

The heat produced in the walls of the vessel is practically negligible. The heat losses to the outside are small, and the heat exchange with the volume of the blood is sufficiently high so as not to cause serious errors. Experiments have shown that the flow pattern is of no influence upon the output if the spacing of the thermoelements from the heating device is more than 1.2 times the diameter of the vessel.

The system has been used for the study of flow rates in a range from about 10 to 300 cm³/min. The output from the thermoelements is in the microvolt range. The rf power requirement at the electrodes is a fraction of 1 watt.

The error varies with the construction and with the flow rate; an average value of the error is $\pm 5$ per cent. Also the dynamic response depends upon the apparatus construction and the flow rate. A sudden variation of the flow rate becomes noticeable within less than 0.2 sec, but the output reaches an equilibrium in about 1 to 5 sec (up to 25 sec). With special systems (improved thermal contact between thermoelement wires and the walls of the vessel), pulsations of the blood stream of about one pulse per second can be observed.

The method has found wide publicity and was, for some time, the only method available for the measurement of blood flow in the unopened vessel. The later-developed induction method (1-62) furnishes more accurate results but requires more space. Both methods require surgical procedures.

## 1-64. SONIC METHOD

The basic setup of a sonic flow-velocity transducer system is shown in Fig. (1-6)6; it consists of a sonic transmitter $T$ and a receiver $R$ separated by a distance $d$. The transmitter sends an acoustic signal (pulse or wave train) into the flowing medium. The signal is received at $R$ at a time $\Delta t$ later. The transit time in the flow direction is $\Delta t_1 = d/(c + v)$, where $c$ is the sound propagation velocity and $v$ the linear flow velocity of the medium. A signal traveling against the flow direction requires a transit time $\Delta t_2 = d/(c - v)$. A sinusoidal signal of the frequency $f$ traveling in the flow direction will arrive with a phase shift $\Delta\varphi_1 = 2\pi fd/(c + v)$; a signal traveling in the opposite direction will have a phase

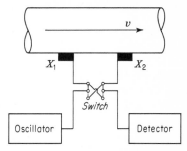

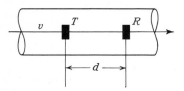

FIG. (1-6)6. Flow-velocity transducer for liquids, sonic method.

FIG. (1-6)7. Modification of the sonic flow-velocity transducer shown in Fig. (1-6)6.

shift $\Delta\varphi_2 = 2\pi fd/(c - v)$. The velocity can be determined from measurements of the transit time or of the phase difference between the emitted and the received signal.

Hess, Swengel, and Waldorf[1] and Kalmus[2] have described sonic systems for the measurement of flow velocities as schematically shown in Fig. (1-6)7. Two piezoelectric crystals $X_1$ and $X_2$ are arranged in the medium or pressed against a plastic tube in which liquid travels, so that sound is transmitted between the crystals and the liquid. An oscillator furnishes an alternating voltage (order of magnitude 100 kc) to the crystal $X_1$. Crystal $X_2$ acts as a microphone. The functions of $X_1$ and $X_2$ are periodically interchanged by a

[1] W. B. Hess, R. C. Swengel, and S. K. Waldorf, *AIEE Misc. Papers* 50–214, August, 1950; *Elec. Eng.*, **69**, 983 (1950); ASME Annual Meeting, *Paper* 54-A-54, 1954; *Elec. Eng.*, **73**, 1082 (1954).

[2] H. P. Kalmus, *Natl. Bur. Standards Tech. News Bull.*, **37**, 30 (1953), and *Rev. Sci. Instr.*, **25**, 201 (1954).

commutator switch. The difference in transit time is measured by a phase-sensitive detector driven synchronously with the commutator switch.

Linear velocities in water as low as 0.1 cm/sec have been measured successfully with this system. The deviation from linearity is indicated to be less than 2 per cent. The dynamic response is limited only by the frequency of the synchronous commutator. The system when applied to the outside of the conduit causes no obstructions or discontinuities in the flow system, but its application is limited to conduits with plastic walls or to large conduits where the transmitter and receiver crystals can be arranged within the conduit. The measurement of flow in metal pipes with an arrangement outside of the pipe is not possible, because of the direct transmission of sound through the metal wall of the pipe, unless special means are used to suppress the influence of this effect.

### 1-65. RADIOACTIVE-TRACER METHODS

Both the electrolytic-tracer method and the electrolytic-dilution method described in 1-61 can be modified by replacing the electrolyte by a radioactive solution and determining the passage of the cloud or the concentration of the liquid with nuclear-radiation detectors.

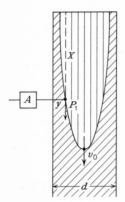

A method suitable for the determination of the velocity with which a liquid runs out of a vessel has been described by Richardson.[1] The method is applicable where the velocity profile has the general form shown in Fig. (1-6)8.

The lower part of a vertical tube is filled with a radioactive-tracer solution, the part above it with a tracer-free solution. As the former runs out and is displaced by the latter, the counting rate measured at a point $A$ outside of the tube decreases.

FIG. (1-6)8. Flow-velocity transducer for liquids, radioactive tracer method.

If $v_p$ is the linear flow velocity at $P$ and $v_0$ the velocity in the center of the conduit, their ratio is

$$\frac{v_p}{v_0} = 4\left[\frac{y}{d} - \left(\frac{y}{d}\right)^2\right] \tag{1}$$

[1] F. M. Richardson et al., *Nucleonics*, **13**, 21 (July, 1955).

The ratio of counting rates at $A$ for circular pipes, according to the authors, is

$$\frac{C_t}{C_0} = 4\left[\frac{y}{d} - \left(\frac{y}{d}\right)^2\right] \tag{2}$$

where $C_t$ is the counting rate at the time $t$ and $C_0$ that at the beginning of the experiments when the lower part of the tube is completely filled with a radioactive-tracer solution.

With an aqueous solution of radioactive sodium 24 and a concentration of 0.05 to 0.16 mc/ml, the authors were able to follow the replacement of the tracer solution by the tracer-free liquid until the thickness $y$ of the layer adjacent to the wall had diminished to 0.002 in.

The application of the method is limited to special cases and single runs.

## 1-66. TRANSDUCERS RESPONDING TO GAS FLOW VELOCITY: THERMAL SYSTEMS

*a. Thomas Method.* The following method originally devised by Thomas[1] has been described primarily for the flow-rate or flow-velocity determination in gases but is, in principle, also applicable to the measurement of liquid flow (see 1-63). The method is illustrated in Fig. (1-6)9. The gas, or liquid, enters a conduit with the temperature $t_1$, which is measured by the thermal transducer $T_1$ (e.g., a resistance thermometer); it then passes a heater $H$ energized by a source $S$, and it emerges with the (higher) temperature $t_2$, which is measured by the resistance thermometer $T_2$. If the specific heat is constant, the flow rate $F$ can be determined from

$$F = \frac{PC}{c_p(t_2 - t_1)}$$

where $P$ is the electric power required to maintain the temperature difference $t_2 - t_1$, $C$ is the mechanic-caloric equivalent, and $c_p$ the specific heat of the gas at constant pressure or the specific heat of the liquid.

The method can be used in connection with a servo system. For this purpose the thermal transducers $T_1$ and $T_2$ are arranged in a bridge circuit, Fig. (1-6)10, which is adjusted so that no current passes through the meter relay $M$ as long as the temperature difference $t_2 - t_1$ is of a given value. A feedback loop controls the current through the heater, and the arrangement acts so as to maintain the

[1] C. C. Thomas, *J. Franklin Inst.*, **172**, 411 (1911).

temperature difference $t_2 - t_1$ constant. The electric power necessary to keep this temperature difference constant is proportional to the flow rate.

The output of the system is an indication of the mass rate of flow (in kilograms per second) rather than the volume rate of flow (in cubic meters per second). A reduction of the reading to standard temperature and pressure is not required, therefore.

The method has been applied primarily to the measurement of large flow rates, as high as 75,000 ft³/hr; the power requirements for

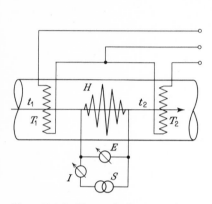

Fig. (1-6)9. Flow-velocity transducer system for gases, thermal method.

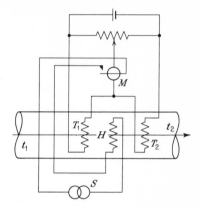

Fig. (1-6)10. Thermal flow-velocity transducer for gases, modification of the system illustrated in Fig. (1-6)9.

such rates are of the order of 1 kW. The restriction of the conduit by the wires is negligible. Turbulence arising on the wires may be considered an advantage because it provides for better mixing.

Errors are likely to arise from nonuniform heating of the gas, from nonuniform distribution of the flow velocity, and from heat loss to the environment (thermal insulation of the conduit is recommended). Deposit of dust on the heater will tend to make the response slower but will not affect the accuracy. Errors resulting from the variation of the specific heat can become large if the composition of the flowing gas is altered. The change of specific heat of gases with temperature and humidity will generally be small and not cause any serious error.

Thermoelements can be used instead of the resistance thermometers.[1] The heat capacity of the thermal transducers, and thus their inertia, can be kept very low so that the thermal system is capable of following pulsations of the gas or liquid stream.

[1] G. W. Penney and C. F. Fecheimer, *J. AIEE*, **47**, 181 (1928).

Goodyear[1] has used thermistors for flow-velocity measurements. He claims that velocities as low as $10^{-3}$ cc/hr can be measured and that an operating range of 1 to $10^5$ can be obtained.

A modification of this method for the measurement of very high and rather localized fluid flow velocities has been described by Cady.[2] The method, called "electronics transit-time anemometer (ETTA)," operates in the following way: A fine wire 0.004 in. in diameter and $\frac{3}{8}$ in. long is stretched, transverse to the flow direction; a current pulse of 40 msec duration is applied that heats the wire and the passing air. Farther downstream is a similar wire acting as a thermal transducer; its resistance variation caused by the passing cloud of hot air

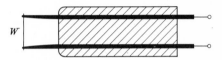

Fig. (1-6)11. Hot-wire probe for flow-velocity measurements in gases.

is measured oscillographically. The transit time should be between 50 $\mu$sec and 6 msec. The method is applicable to flow velocities from 100 to 5,000 ft/sec. The accuracy is of the order $\pm 2$ per cent.

A similar method, described by Walker and Westenberger,[3] permits the measurement of gas velocity in the range of 40 to 400 cm/sec with an error of less than 2 per cent.

*b. Hot-wire Probe.* The basic arrangement is shown in Fig. (1-6)11 and consists of a fine wire $W$ of a material with a large temperature coefficient of resistivity, such as nickel or platinum  The wire is heated by an electric current and cooled by the gas stream. The heat loss causes a decrease of the wire temperature and thus of the wire resistance. The rate of heat loss depends also upon the physical characteristics of the medium and upon the geometry and the surface structure of the wire.

The system can be used in two ways, either the heater voltage or the heater current is kept constant, and the resistance variations of the wire are used as a measure for the flow velocity of the gas in the vicinity of the probe. The transfer function, resistance variation versus gas flow velocity, increases first steeply and, at higher velocities, with gradually decreasing slope. The method is suitable for the measurement of small velocities. Alternatively, the resistance of the wire, i.e., its temperature, is kept constant by varying the

[1] R. S. Goodyear, *Elec. Mfg.*, October, 1956, p. 90.

[2] W. M. Cady, Physical Measurements in Gas Dynamics and Combustion, in R. W. Ladenburg (ed.), "High Speed Aerodynamics and Jet Propulsion," vol. 9, sec. C, 3, part I, Princeton University Press, Princeton, N.J., 1954.

[3] R. E. Walker and A. A. Westenberger, *Rev. Sci. Instr.*, **27**, 844 (1956).

heater current or the voltage, and the current or voltage variations are used as a measure of the gas flow velocity. The heat loss, and hence the square of the heater current $I$, is approximately proportional to the square root of the gas velocity $v$.[1]

$$I^2 = B\sqrt{v} + C$$

The resistance-variation method is usually preferred, although the sensitivity obtained by the constant-temperature method is superior.

The method is applicable for flow velocities ranging from 0.5 cm/sec up to very high velocities in the supersonic range. The application of the method in subsonic and supersonic flow is extensively discussed by Kovásznay.[2]

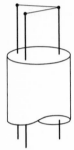

Temperature variations of the gas will cause changes in the transducer output; the temperature influence can be reduced considerably and readings practically independent of the gas temperature can be obtained if the temperature of the wire is high. Another means to reduce the influence of gas-temperature variations upon the transducer output consists of a bridge circuit in which one wire is exposed to the flowing gas and a wire in an adjacent bridge arm is exposed to the resting gas by mounting it in a shielding tube in the gas stream.[3]

FIG. (1-6)12. Direction-sensitive hot-wire probe.

The response varies as the position of the probe changes with respect to the flow direction; it is a maximum if the wire is perpendicular to the flow direction. Arrangements with two wires set under an angle permit the identification of the angle of flow. Figure (1-6)12 shows an arrangement for measuring airflow direction with two wires set under an angle of about 10°.[4] The same authors also use three- and four-wire arrangements for three-dimensional direction analysis and temperature-independent flow measurement. Fay[5] described an arrangement with six wires which measures velocity

[1] L. V. King, *Phil. Trans. Roy. Soc. (London)*, (A) **214**, 373 (1914); see also J. M. Burgers, in W. Wien-F. Harms (eds.), "Handbuch der Experimentalphysik," 4th ed., part I, p. 637, Akadem. Verl. Ges. m.b.H., Leipzig, 1931.

[2] L. S. G. Kovásznay, Physical Measurements in Gas Dynamics and Combustion, in R. W. Ladenburg (ed.), "High Speed Aerodynamics and Jet Propulsion," vol. 9, sec. F2, Princeton University Press, Princeton, N.J., 1954.

[3] G. S. C. Thomas, *Phil. Mag.*, (6) **39**, 505 (1920).

[4] L. F. G. Simmons and A. Bailey, *Phil. Mag.*, (7) **3**, 81 (1927).

[5] R. D. Fay, *J. Franklin Inst.*, **183**, 785 (1917).

fluctuation and cancels out temperature-fluctuation effects. Directional-sensitive hot-wire probes can also be made by the use of two wires parallel to each other and spaced at a distance of about 0.5 mm.[1]

When measuring at very low air speeds, the hot-wire probe must be used in the position in which it was calibrated, since the self-induced air current varies with the probe position.

The transducer is apparently not (or very little) influenced by humidity and, at least at steady flow and low velocities, by the pressure of the medium. But the temperature of the conduit wall can be of disturbing influence if the hot wire is in close proximity to the wall (order of 2 mm). This influence is reduced at high gas velocities. Errors are also caused by the deposit of dirt or dust upon the wire, by mechanical stress in the wire, and by vibration of the suspension wires under the influence of the gas-stream impact. Further errors can be caused by a variation of the composition of the gas mixture.

The finer the wire, the faster it will follow a fast variation in flow velocity, but the easier it will break or will be altered by mechanical stress. Platinum Wollaston wires of a thickness of the order of 1 $\mu$ respond in general to gas-velocity fluctuations in the frequency range of 1,000 to 10,000 cps (in special cases up to 100,000 cps). The response of a hot-wire probe to alternating air currents has been studied by Richards[2] and Maxwell[3] by mounting the hot-wire transducers on a tuning fork. The time response of a hot-wire system can also be measured, according to Ziegler,[4] by connecting the hot-wire transducer in a bridge and applying a direct current and a superimposed alternating current or square wave to the bridge input. The instantaneous output of the bridge is proportional to the resistance fluctuation of the wire and the resultant signal can be used for the determination of the frequency response.

Hot-wire probes are usually made from platinum, platinum 80-irridium 20 alloy, tungsten, or nickel. The length of the wires varies from several centimeters to a fraction of a millimeter; the diameter varies from 0.1 mm down to the micron range (0.00005 in.).

Heated thermoelements can be used instead of heated wires, but their time constant is considerably larger than that of a single hot wire, and their output depends strongly upon heating-current

[1] Thomas, *loc. cit.*

[2] R. C. Richards, *Phil. Mag.*, (6) **45**, 926 (1923).

[3] R. S. Maxwell, *Phil. Mag.*, (7) **6**, 945 (1928).

[4] M. Ziegler, *Proc. Koninkl. Ned. Akad. Wetenschap.*, **34**, 663 (1931).

variations (at a flow velocity of 25 cm/sec, a change of 1.5 per cent heating current causes a reading error of $\pm 10$ per cent).

The use of thermistors for flow-velocity measurements has been investigated by Wachter.[1] As expected, the calibration curve depends upon the geometrical shape of the thermistor.

The hot-wire method is the only method which permits the local measurement of velocity in a small volume and which allows for the instantaneous measurement of velocity and velocity fluctuations in unsteady flow. It can be used in flow fields with high gradients and fast variations of velocity, but the hot wire presents an obstruction to a flow system which may not always be negligible. The output depends upon the specific heat of the gas. In gas mixtures, errors are likely to occur if the ratio of the components is altered, unless bridge circuits are used to balance out this effect. Because of variation in surface structure, wires have to be calibrated individually. Also wires are likely to change as the result of structural crystallographic variations, mechanical stress, deposition of dust and, at high temperatures, evaporation. The nonlinearity of the wire can be compensated for by the use of elements with square characteristics in subsequent stages. Methods for such linearizations can be found in Jacobs.[2]

For summarizing reviews and references, see Kovásznay, *loc. cit.*, and Jacobs, *loc. cit.* For the measurement of shock waves in gases, see D. S. Dosanjh, *Rev. Sci. Instr.*, **26**, 65 (1955), and J. G. Clouston, L. J. Drummond, and W. F. Hunter, *J. Sci. Instr.*, **34**, 321 (1957). For an application of the same principle for acoustic measurements ("hot-wire microphone") see W. S. Tucker and E. T. Paris, *Phil. Trans. Roy. Soc.* (*London*), (A) **221**, 389 (1920–1921).

1-67. SONIC SYSTEMS

The sonic gas-flow-velocity transducer is based upon the propagation velocity of sound which depends upon the flow velocity of the medium in which propagation takes place. The principle is identical with that explained in 1-64 for liquid flow-velocity transducers.

An anemometer with four sound receivers oriented at the cardinal points of the compass in a 10-ft-diameter circle around a sound head is described by Corby.[3] The time difference between the arrival of the signals at opposite receivers is 15 $\mu$sec/mph of wind velocity. The instrument displays the instantaneous wind velocity and

[1] H. Wachter, *Arch. tech. Messen*, V 116–5, September, 1956.

[2] U. Jacobs, *Arch. tech. Messen*, V 116–3, January, 1954, and V 116–4, February, 1954.

[3] R. E. Corby, *Electronics*, **23**, 88 (January, 1950).

direction as a vector on the screen of a cathode-ray tube and is accurate within 0.5 per cent for wind velocities of 50 mph.

### 1-68. TRANSPORTATION-OF-CHARGE METHODS

This method is illustrated in Fig. (1-6)13. Ions are produced at a place $P_1$ in the flow duct; the ion cloud follows the air with great

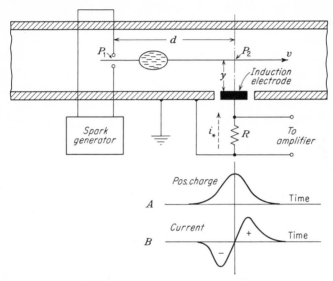

FIG. (1-6)13. Flow-velocity transducer system based upon the transportation-of-charge method.

fidelity without influencing the flow pattern. The presence of ions is detected at a place $P_2$ downstream at a time $\Delta t$ later. The flow velocity can be determined from

$$v = \frac{d}{\Delta t}$$

where $d$ is the distance between $P_1$ and $P_2$.

The method measures the average stream velocity of the moving-ion volume element. It cannot give information about the localized or instantaneous velocity, and it cannot measure the total average velocity.

The ions are mostly produced within the gas stream by a high-voltage spark discharge between two electrodes, the voltage being generated either by an induction coil or by a pulse-modulated rf source. The latter method is preferable; it permits the control of the pulse length. In air (oxygen) the spark method furnishes ionization predominantly of a negative sign and strong enough for convenient measurements; but the electrodes projecting into the gas-flow conduit are likely to alter the flow pattern. Ionization by α particles (modulated by rotating or oscillating shutters), by high-energy electron beams, soft X rays, ultraviolet radiation, and corona discharges have been considered or tried but have apparently not led to practical results.

Two different types of detectors are used, the ion collector, i.e., an electrode that is charged by a supply voltage and discharged by the ions which it collects, and the induction electrode pickup, in which, through electrostatic induction, a voltage is induced by the passing ion cloud. Both types can be mounted either protruding into the gas stream or flush with the wall. Figure (1-6)14 shows an ion

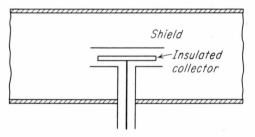

Fig. (1-6)14. Ion collector for flow-velocity measurements in gases.

collector protruding into the gas conduit. It is initially charged to a high potential and discharged by the ions it attracts. Its main disadvantage is the fact that it imparts a velocity to the ions which is different from that of the neutral gas molecules. The output signal from such a collector (when connected to an amplifier with 10 MΩ input impedance) is of the order of 1 mV. The duration of the pulse is about 0.1 msec. The flush-mounted induction pickup is generally preferred, its influence upon the flow pattern is negligible, and its output signal permits accurate definition of the passage of the ion cloud over the pickup electrode. It is insensitive, however, to ion clouds having exactly balanced positive and negative charge distribution. The electrode can be formed either by a section of the pipe or by a metal plate or button set flush with the wall and insulated from the rest of the pipe, as shown in Fig. (1-6)13.

The signal definition depends upon the width of the electrode in the flow direction. For the cylindrical electrode, a width of 20 to 30 per cent of the tube diameter is recommended; for the flush-mounted electrode the width should be about one-half the approach distance, or $y$ in Fig. (1-6)13.

A theoretical treatment of the induction process is given by Shockley[1] and by Bouwcamp and de Brujin.[2] The polarity of the peak charge and of the current is determined by the sign of the charge of the ion cloud. The approach of a predominantly negative charge causes the flow of positive charges from ground to the pickup electrode $P$ (arrow). The flow direction of positive charges is from the positive to the negative pole; hence the grid of the following amplifier becomes negative during the approach of the negative cloud, and positive thereafter, as shown in curve $B$, Fig. (1-6)13. The cloud may travel as much as 2 to 3 in. distance from the electrode. The reference point for the determination of the time difference $\Delta t$ is measured from the point of maximum charge or of zero current and can be determined, in the latter case, to within 0.2 msec. The ion cloud will diffuse as it moves downstream, but for drift distances up to 12 in., the ionization will remain localized within a cylinder of $1\frac{1}{4}$ in. diameter at a drift velocity of 150 mph. Moderate lengthening of the discharge time, that is, of the time necessary to produce the ion cloud, does not change the waveform at the pickup electrode appreciably, compared to a point charge.

Gas velocities as low as 20 mph have been measured with this method. The method is not applicable for smaller velocities (because of radial and axial diffusion and of recombination of the ions) but has been used successfully in supersonic flow up to Mach 2 (twice the velocity of sound). Path lengths in the order of 1 to 10 in. have been used. The accuracy of the method at high velocity is difficult to evaluate because precision methods for comparison do not exist; at best the error is estimated with a few per cent. Semicontinuous operation can be obtained by triggering the gas discharge from the pickup electrode via a feedback loop and observing the recurrence frequency, which depends upon the ion transit time. A system of this kind is described by Mellen.[3]

For references see W. M. Cady, Physical Measurements in Gas Dynamics and Combustion, in R. W. Ladenburg (ed.), "High Speed Aerodynamics and Jet Propulsion," vol. 9, Princeton University Press, Princeton, N.J., 1954.

[1] W. Shockley, *J. Appl. Physics*, **9**, 635 (1938).
[2] C. F. Bouwcamp and N. G. de Brujin, *J. Appl. Physics*, **17**, 562 (1945), and Errata, **19**, 105 (1947).
[3] G. L. Mellen, *Electronics*, **23**, 80 (February, 1950).

## 1-7. Humidity Transducers

The field is divided into two groups: systems for humidity determination in gases (1-71 and 1-72) and systems for measurements in liquids or solids (1-73).

One group of direct-acting transducer systems that furnish an electric output in response to the humidity of a surrounding gas atmosphere is based upon the absorption of humidity in a probe, causing dissociation of molecules into ions. Electrolytic conduction takes place; the resistance of the probe is a function of the humidity (1-71*a*). An empirical calibration is usually required. However, two methods have been described which permit the absolute determination of the absorbed water in the probe (1-71*b* and *c*).

A second group of transducers is based upon the variation of the dielectric constant of a gas–water vapor mixture (1-72). The dielectric constant of dry gases is usually not much different from 1, while that of water is much higher. Thus, the concentration of water vapor in a gas may be detected from the measurement of capacitance (1-72*a*) or the resonance frequency of a cavity (1-72*b*).

The humidity in gases can also be determined by measuring the thermal conductivity of the gas–water vapor mixture,[1] the mobility of ions,[2] or the absorption of nuclear radiation (see Radiation Thickness Gauges, 1-16). A method for the measurement of very small degrees of relative humidity (0 to 2 per cent) in air at reduced pressure is described by Hinzpeter and Meier.[3] The method is based on the variation with humidity of the cathode fall and the breakdown voltage in electric gas discharges and is primarily applicable to humidity determination during freeze drying.

There exist also a number of indirect systems for the determination of humidity in gases. One of these is the wet-and-dry bulb psychrometer with electrical readout (the liquid thermometer being replaced by resistance thermometers, thermistors, or thermoelements). Information on such systems may be found in Guthmann;[4] another indirect method, the dew-point hygrometer with electrical readout, is described in a reviewing paper by Czepek.[5] The formation of the

---

[1] C. Z. Rosecrans, *Ind. Eng. Chem. Anal. Ed.*, **2**, 129 (1930).

[2] E. Griffiths and J. H. Awbery, *Proc. Physic. Soc. (London)*, **41**, 240 (1928).

[3] A. R. Hinzpeter and W. Meier, *Z. angew. Physik*, **3**, 216 (1951).

[4] K. Guthmann, *Arch. tech. Messen*, V 1283–3, September, 1932, and V 1283–11, October, 1953; also *Arch. tech. Messen*, V 1283–2, 1932, and V 1283–4, July, 1933.

[5] R. Czepek, *Arch. tech. Messen*, V 1283–6, August, 1940; see also C. W. Thornthwaite and J. C. Owen, *Monthly Weather Rev.*, **68**, 315 (1940); C. W. Sisco, *Instrumentation*, **2**, 14 (July–August, 1947), and Guthmann, *loc. cit.*

condensation of water vapor on a cooled mirror surface causes a change of light reflection, which is detected photoelectrically, or the condensation changes the resistance between two metal electrodes.

Other indirect systems have been constructed from hygrometers made of dimensionally varying materials, such as the hair hygrometer in combination with displacement transducers. Some systems of this type are reviewed by Wexler.[1] Humidity can also be found from optical measurements, e.g., infrared absorption in connection with photoelectric devices.

The moisture content in dielectric liquids and solids changes their dielectric properties, such as the resistivity and the dielectric constant. Methods for moisture determination in solids, based upon the variation with humidity of these electric properties, are described in 1-73*a*. Another method for moisture determination by means of nuclear magnetic resonance measurements is described in 1-73*b*.

The humidity in solids can sometimes be determined by forming a closed gas space adjacent to the solid. The gas is allowed to come into humidity equilibrium with the solid, and the humidity of the gas is then measured. In some cases, humidity in solids can be determined by means of the beta gauge (1-16).

For general information see A. Wexler and W. B. Brombacher, Methods of Measuring Humidity and Testing Hygrometers, *Natl. Bur. Standards Circ.* 512, September, 1951; Wexler, *loc. cit.*

### 1-71. RESISTIVE TRANSDUCERS RESPONDING TO HUMIDITY IN GASES

*a. Resistance-variation Systems.* Most resistive humidity transducers consist essentially of a hygroscopic body which absorbs water from the surrounding atmosphere and which contains a salt that dissociates in the presence of water. The concentration of dissociated ions is determined by the measurement of the electric resistance between electrodes applied to the test specimen.

An example of a resistive humidity transducer is shown in Fig. (1-7)1. An insulating plate carries two strips of metal; the space between the strips is coated with a thin layer of a hygroscopic material. Figure (1-7)2 shows a similar transducer; the arrangement with the interdigitating metal strips has the effect of reducing the resistance between the metal electrodes. Moreover, small local injuries of the resistive layer do not greatly affect the results. A picture of a commercial humidity transducer of this type is shown in Fig. (1-7)3.

[1] A. Wexler, Electric Hygrometers, *Natl. Bur. Standards Circ.* 586, sec. 5, p. 13, September, 1957.

A considerable number of variations of the basic system have been described in the literature.[1] The hygroscopic material is either deposited on impervious insulating surfaces, on fibers or fabrics, or on porous ceramics. A great number of hygroscopic materials have

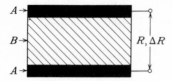

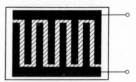

Fig. (1-7)1. Resistive humidity transducer; the hygroscopic salt layer $B$ is deposited between the two electrodes $AA$.

Fig. (1-7)2. Electrode arrangement of a resistive humidity transducer.

Fig. (1-7)3. Resistive humidity transducer (*courtesy of El-Tronics, Inc., Mayfield, Pa.*).

also been described or suggested, e.g., sulfuric acid, phosphoric acid, lithium chloride, calcium chloride, zinc chloride, solutions of the tetrachlorides of the metals tin, zirconium, hafnium, or lead, and many others.

The resistance of a sensing element of the described type varies for a humidity variation from zero to 100 per cent by large orders of

---

[1] Wexler, *loc. cit.*

magnitudes, e.g., from $10^8$ to $10^4$ ohms. This fact makes it sometimes difficult to find an electric system capable of measuring and indicating the entire humidity range on one scale. The measuring current should be small so as not to cause an undue heating of the probe.

Several other difficulties are likely to arise in the construction and the use of such transducers. The effect of water absorption as well as that of dissociation depends strongly upon temperature. The humidity-temperature relationship of different salts has been studied by Wexler and Hasegawa.[1]

If the hygroscopic body is thick, it will take a long time for it to come to equilibrium with the surrounding atmosphere; thus the dynamic response is poor and the indication of the system will show a lag, sometimes of several hours. If the hygroscopic body is thin, so that the physical process is limited to the surface or near-surface layer, the system is easily affected by surface contamination and mechanical injury, and the measuring results may become nonreproducible. The sensing elements should be protected, therefore, from exposure to 100 per cent humidity, because the condensation water is likely to damage the vulnerable surface.

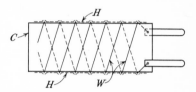

Fig. (1-7)4. Resistive humidity transducer; $C$, insulating carrier; $W$, wires; $H$, hygroscopic coating.

However, Verzár, Keith, and Parchet[2] have used a 2 per cent solution of sodium polyacrylate. Transducers made with this material are resistant to the formation of water droplets; their highest sensitivity is in the range between 80 and 100 per cent relative humidity.

Probably the most successful humidity transducer as far as accuracy, speed of response, and stability are concerned has been described by F. W. Dunmore and coworkers.

[1] H. Diamond, W. S. Hinman, Jr., and F. W. Dunmore, *J. Research Natl. Bur. Standards*, **20**, 369 (1938); F. W. Dunmore, *J. Research Natl. Bur. Standards*, **20**, 723 (1938); F. W. Dunmore, *Bull. Am. Meteorol. Soc.*, **19** (6), 225 (1938); F. W. Dunmore, *J. Research Natl. Bur. Standards*, **23**, 701 (1939); H. Diamond, W. S. Hinman, F. W. Dunmore, and E. G. Lapham, *J. Research Natl. Bur. Standards*, **25**, 327 (1940).

The instrument is illustrated in Fig. (1-7)4. It contains an insulating carrier $C$ (polystyrene) supporting two wires $W$ of noble metal

[1] A. Wexler and S. Hasegawa, *J. Research Natl. Bur. Standards*, **53**, 19 (1954).
[2] F. Verzár, J. Keith, and V. Parchet, *Arch. ges. Physiol.*, *Pflügers*, **257**, 400 (1953).

which are wound in the form of two parallel helices. A hygroscopic coating $H$ consisting of polyvinyl acetate or polyvinyl alcohol with a dilute lithium chloride solution is applied to the surface of the insulating carrier and makes contact with the wires. The resistance between the wires is measured with an a-c method to avoid polarization. The resistance decreases with increased humidity in a function approximately proportional with the logarithm of the relative humidity. A calibration characteristic is shown in Fig. (1-7)5.

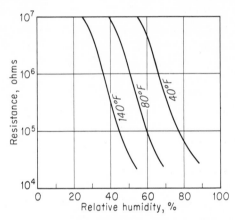

FIG. (1-7)5. Calibration curves of the humidity transducer illustrated in Fig. (1-7)4.

In a practical sense, the useful resistance variation of such a transducer element extends only over a limited range of relative humidity, e.g., from 20 to 35 per cent; transducer elements for different ranges can be made by using different concentrations of the electrolyte in the hygroscopic layer. Several transducer elements are needed to cover the total range from 2 to 99 per cent of relative humidity. These transducers must either be exchanged for the measurements in different ranges or used in a resistive network to cover the entire range.

The resistance of a single transducer varies from about $10^4$ to $10^7$ ohms. The resistance at constant humidity depends strongly upon the temperature, so that either operation at constant temperature or temperature effect compensation is required. A method of temperature compensation is shown by Glückauf.[1]

Transducers of the described type are accurate and reproducible to within $\pm 2$ to $\pm 2.5$ per cent; special sensing elements with an

[1] E. Glückauf, *Proc. Phys. Soc. (London)*, **59**, 357 (1947).

error that does not exceed $\pm 1.5$ per cent are commercially available. Variations of relative humidity as small as 0.15 per cent can be detected. The calibration curve is nonlinear; empirical calibration is required.

The speed of response is high. If exposed to an abrupt, large change of humidity at room temperature, the element will assume 63 per cent of its equilibrium resistance within several seconds. But the speed of response depends upon the velocity of the air current.[1] It also depends upon the magnitude of the humidity change and upon the temperature. The resistance changes faster when the humidity varies from a low to a higher value than it does in the opposite direction. The response at low temperature is slower than at high temperature.

Transducer systems with very fast response have been described by Wexler and Krinsky.[2] Such systems consist of a thin layer of potassium dihydrogen phosphate, which is deposited on glass by evaporation in vacuum and which is converted into potassium metaphosphate. If exposed to a humidity change from 83.3 to 100 per cent, these transducers will reach 63 per cent of the final resistance in less than $\frac{1}{2}$ sec at room temperature and in about 10 sec at $-20°C$. The response time of this transducer is decreased by a factor of about 10 to 30 as compared to the lithium chloride transducer.

An electrical hygrometer based upon the variations with humidity of the electrical properties of anodized aluminium oxide layers has been suggested by Ansbacher and has been investigated and described by Cutting, Jason, and Wood.[3]

The transducer consists essentially of a capacitor having as a dielectric a porous film of aluminum oxide. At least one capacitor plate should be permeable for the water vapor to permit the penetration of humidity to the aluminum oxide. Both the capacitance and the resistance of such a capacitor change in response to the ambient humidity.

The porous oxide layer is formed on aluminum by anodization in acid electrolyte (17.5 volume per cent $H_2SO_4$, current density 10 to 100 mA/cm$^2$, 30 min). The apparent dielectric constant of such a

[1] Glückauf, *ibid.*, p. 344.

[2] A. Wexler and A. Krinsky, *Natl. Bur. Standards Tech. News Bull.*, June, 1954; also A. Wexler et al., *J. Research Natl. Bur. Standards*, **55**, 71 (1955).

[3] C. L. Cutting, A. C. Jason, and J. L. Wood, *J. Sci. Instr.*, **32**, 425 (1955); see also C. R. Underwood and R. C. Honslip, *J. Sci. Instr.*, **32**, 432 (1955). The variation with humidity of the dielectrical properties of aluminum oxide has been observed formerly by A. V. Astin, *J. Research Natl. Bur. Standards*, **22**, 690 (1939).

film in the vicinity of the saturation point often exceeds 1,000; apparent dielectric constants as high as 8,000 have been observed. The effect is explained by assuming a structure of the aluminum oxide layer like that shown in Fig. (1-7)6. The deposition of water on the inside walls of the tubular pores leads to a reduction of the resistance $R_1$ and to an apparent increase of capacitance between the applied electrodes. At high degrees of humidity, the capacitance is

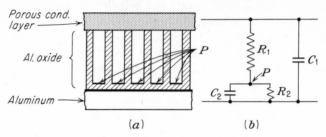

(a)                    (b)

FIG. (1-7)6. Humidity transducer based upon the variations with humidity of the dielectric properties of aluminum oxide: (a) schematic diagram; (b) equivalent circuit [*from C. L. Cutting, A. C. Jason, and J. L. Wood, J. Sci. Instr.*, **32**, 425 (1955); *by permission*].

essentially that between the pore bases $P$ and the aluminum on which the oxide layer is formed.

A practical form of the transducer is shown in Fig. (1-7)7. It consists of an aluminum rod which is oxidized over a part of its surface.

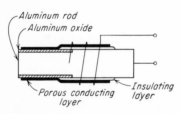

FIG. (1-7)7. Construction of an aluminum oxide humidity transducer [*C. L. Cutting, A. C. Jason, and J. L. Wood, J. Sci. Instr.*, **32**, 425 (1955); *by permission*].

The rod serves as one electrode of the probe; a porous conducting layer (graphite or a thin evaporated metal layer) applied to the outside of the alumina serves as the other electrode. The probe behaves electrically like a parallel combination of a resistance and a capacitance. A typical calibration curve is shown in Fig. (1-7)8. The transducer can be used for a range of relative humidity from 0 to 100 per cent. The sensitivity (slope of the characteristic) decreases near the saturation point.

The error of the method is estimated to be about 3 per cent. The dynamic response to a sudden variation of ambient humidity depends upon the magnitude and direction of such a variation. Equilibrium is established, on the average, in 10 to 100 sec.  The response at humidities higher than 80 per cent is slower, and also a slight hysteresis effect is noticeable at high degrees of relative humidity.  An aging effect occurs during the first few months after the formation of

the oxide layer, resulting in a reduction of the capacitance and an increase of resistance at any given humidity.

The transducer output is a function of the ambient relative humidity only.  In the temperature range between $-15$ and $+80°C$, the readings are practically independent of temperature. Changes of capacitance and resistance with variation of ambient relative humidity can be observed at temperatures above $100°C$ (even as high as $400°C$), but the exposure to such high temperatures can cause an irreversible change of the characteristic.

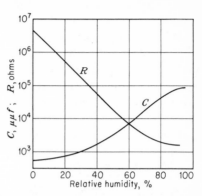

FIG. (1-7)8. Capacitance and resistance of an aluminum oxide humidity transducer plotted against the relative humidity of the surrounding atmosphere [*C. L. Cutting, A. C. Jason, and J. L. Wood, J. Sci. Instr.*, **32,** 425 (1955); *by permission*].

*b. Thermal System (Dewcel[1]).*  Although similar in appearance to the preceding methods, this system operates on a different principle and permits an absolute determination of humidity.  The element is illustrated in Fig. (1-7)9.  A tubular wick made from glass fibers and

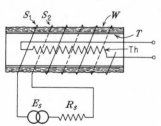

FIG. (1-7)9.  Humidity transducer, thermal system.

impregnated with a hygroscopic salt (lithium chloride) is mounted over a thin-walled metal tube $T$ electrically insulated from the tube.  Two parallel silver wires $S_1$ and $S_2$ are helically wound around the wick and connected to an alternating-voltage source $E_s$ through a current-limiting resistor $R_s$.  The lithium chloride absorbs humidity from the surrounding air and becomes conductive, so that a current passes from one silver wire through the lithium chloride layer to the other silver wire.  The current generates heat in the lithium chloride layer and tends to evaporate the humidity from it, a process which leads to an increase of resistance and a reduction of the current.  An equilibrium

[1] Trade name, Foxboro Company, Foxboro, Mass.

is reached when the layer neither gains from nor loses water to the surrounding atmosphere; i.e., equilibrium is reached at that temperature of the lithium chloride layer at which the partial pressure of water over a saturated lithium solution just equals the ambient water-vapor pressure. The temperature is measured by means of a resistance thermometer $Th$ (or with any other thermometer), which should be in good thermal contact with the lithium chloride layer.

The vapor pressure of saturated lithium chloride solution at different temperatures is accurately known from tables; the instrument needs no empirical calibration, except that for the resistance thermometer. The output can be calibrated in dew-point temperature directly. The instrument can be used over a range of relative humidities from about 15 to 100 per cent at temperatures between 5 and

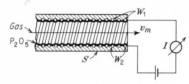

Fig. (1-7)10. Humidity transducer, electrolysis system, schematic diagram.

100°C and a range of dew-point temperatures from about −30° to about +70°C. The upper limit is imposed by the thermal instability of the insulating materials used in the construction; the lower limit by the fact that the lithium chloride layer cannot reach a temperature lower than the ambient temperature. Therefore, the water-vapor pressure over saturated lithium chloride solution at ambient temperature constitutes the lowest limit of the range of this method.

The error of the instrument, expressed in dew-point temperature, is of the order of $\pm 2$ to 3°F. The time required to reach 98 per cent of the equilibrium temperature is 2 to 4 min. The system must be shielded against rapid changes of wind speed and against water droplets. Reconditioning of the cell (with lithium chloride) every 90 to 100 days is recommended.

For literature see W. F. Hickes, *Refrig. Eng.*, **54**, 351 (1947), and *Instruments*, **20**, 1128 (1947); also J. H. Conover, *Bull. Am. Meteorol. Soc.*, **31**, 13 (1950).

*c. Electrolysis System.* The system is illustrated schematically in Fig. (1-7)10. The sensing element $S$ consists of two platinum wires $W_1$ and $W_2$ helically wound inside a Teflon tube of 0.02 in. ID. The space between the wires is coated with a thin layer of partially hydrated phosphorous pentoxide ($P_2O_5$). A gas stream flows continuously through the tube; the humidity in the gas is absorbed by the $P_2O_5$ layer. A d-c voltage which is large compared to the polarization voltage ($\gg 2$ volts) is applied to the wires; the water is decomposed by electrolysis into gaseous $H_2$ and $O_2$. The current $I$ during the time $t$ will decompose a mass of water of

$$m_{water} = AIt$$

If a humid gas flows continuously through the tube with a constant-mass flow velocity $v_m = m_{mixture}/t \approx m_{gas}/t$, the system will come to an equilibrium when the rate of water absorbed by the $P_2O_5$ layer is equal to that electrolytically decomposed. The current is then

$$I = k \frac{V_w}{V_g} v_m \tag{1}$$

where $V_w$ and $V_g$ are the partial volumina of water and gas. The constant $k$ is

$$k = \frac{2 \times 96,500}{18.016} \times \frac{W_{water}}{W_{gas}}$$

where $W_{water}$ and $W_{gas}$ are the gram-molecular weights of water and of the gas. In the case of water in air, Eq. (1) will take the form

$$I = 0.0066 \times cv_m$$

where $c$ $(= V_w/V_{air})$ is the concentration by volume of water in air, expressed in parts per million (ppm). If the volume flow velocity is 100 cc/min, the current will be 13.2 $\mu A$/ppm by volume (at a temperature of 25°C and a pressure of 1 atm).

The instrument is particularly useful for very small concentrations of water in the gas, between 1 and 1,000 ppm; commercial instruments (e.g., Beckman Hygrometer, Fullerton, Calif.) have in general ranges from 0–10 to 0–1,000 ppm. The lower limit is about 1 ppm.

The accuracy of the system is between $\pm3$ to $\pm5$ per cent of the full scale reading. In the lower ranges of humidity, the water in the flowing gas will not come into equilibrium with that in the $P_2O_5$ layer. The "efficiency" of the absorption depends upon the flow rate and the humidity in the gas. Correction curves for freon 12 are published by Taylor.[1] Because of the incomplete absorption, empirical calibration is frequently required.

The dynamic response of the instrument is moderately high. If the humidity is suddenly increased, the instrument will reach 63 per cent of the final indication within 1 min. The time lag for a decrease of humidity is about 2 min.

The pressure of the gas mixture and its mass flow velocity must be kept constant. If the *volume* flow velocity is kept constant, as it is in

[1] E. S. Taylor, *Refrig. Eng.*, **64**, 41 (July, 1956).

most practical cases, the temperature of the gas mixture must also be kept constant; otherwise a variation of the temperature causes an error of 0.3 per cent/deg. The instrument cannot be used for gases that interact with phosphorous pentoxide.

### 1-72. DIELECTRIC SYSTEMS

*a. Capacitive System.* The high dielectric constant of water ($\sim$80) suggests the use of capacitive methods and transducers for the determination of humidity in gases. However, the low concentration of water in air, even at saturation, causes only a small variation of capacitance. The dielectric constant of gases under normal conditions changes with a variation of water content only in the fourth and fifth decimal position (e.g., dielectric constant of dry air at 45°C is 1.000247; at saturation it is 1.000593). Therefore, considerable experimental means are required for the detection and measurement of such small capacitance variations.

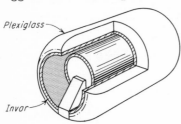

FIG. (1-7)11. Capacitive humidity transducer [*from P. Webb and M. K. Neugebauer, Rev. Sci. Instr.*, **25**, 1212 (1954); *by permission*].

A capacitive system for the determination of humidity in air has been described by Webb and Neugebauer.[1] The air flows through a capacitor, illustrated in Fig. (1-7)11, consisting of two concentric cylinders. The capacitor is thermally insulated on the outside and slightly heated. The change of capacitance in response to the variation of water vapor in the air stream causes a frequency variation of an oscillator operated at about 2 Mc, which is measured with a beat-frequency method. The system detects changes of 1 mg water vapor per liter of air. The dynamic response is fast and limited only by the flushing time; 50 per cent of the final response is obtained in 0.1 sec.

*b. Operation at Microwave Frequencies (Microwave Refractometer).* The most accurate method for the determination of moisture in gases is the microwave refractometer. An instrument of this type is described by Sargent.[2] The method is based upon earlier work.[3]

The system is essentially an arrangement in which the resonance

[1] P. Webb and M. K. Neugebauer, *Rev. Sci. Instr.*, **25**, 1212 (1954).

[2] J. Sargent, *Natl. Bur. Standards Rept.* 4257, June, 1955.

[3] G. Birnbaum, S. J. Kryder, and H. Lyons, *J. Appl. Physics*, **22**, 95 (1951); G. Birnbaum and S. K. Chatterje, *J. Appl. Physics*, **23**, 220 (1952); H. E. Bussey and G. Birnbaum, *J. Research Natl. Bur. Standards*, **51**, 171 (1953).

frequency of a cavity is measured. The resonance frequency varies with the dielectric constant of the material contained in the cavity. If $f_0$ is the resonance frequency of the cavity when it is filled with a reference gas (dry gas, dielectric constant $\epsilon_0$) and $f_1$ the resonance frequency when the cavity is filled with the gas to be examined (dielectric constant $\epsilon_1$), then

$$\frac{\epsilon_1}{\epsilon_0} = \left(\frac{f_1}{f_0}\right)^2$$

The quantity $\epsilon_1$ is the complex dielectric constant; however, preliminary experiments have shown that the loss factor for gases and water vapor can be neglected in the region of $10^{10}$ cps.

A schematic diagram of the system is shown in Fig. (1-7)12. Two resonant cavities are excited by a microwave source (klystron) $K$.

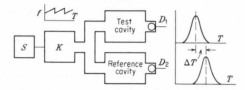

FIG. (1-7)12. Microwave refractometer for humidity determination in gases; schematic diagram.

The resonance in each cavity is indicated by the output from a crystal diode $D$. A frequency modulator $S$ shifts the klystron frequency in a saw-tooth fashion. If the two gases in the cavities have different dielectric constants, resonance will first occur in one cavity and afterwards in the other cavity, as indicated by the curves on the right side of Fig. (1-7)12. The resonance maximum appearing at the crystal detectors will be displaced in time by an amount which is proportional to the difference between the resonance frequencies of the cavities. At an operating frequency of $10^{10}$ cps the difference in resonance frequency, if one cavity is filled with dry air and the other cavity with air of a vapor pressure of 100 mb, is 0.27 Mc. The time difference is measured electronically. In the final form of the instrument a null-balance technique is used; the output is applied to a servosystem which mechanically changes the resonance frequency (volume) of the reference cavity.

The gas temperature and the pressure in either cavity are kept constant; therefore, the output scale can be calibrated directly in vapor pressure.

Tests of the instrument at high vapor pressure, i.e., in the range from 4 to 16 mb vapor pressure, revealed an error in terms of vapor pressure of 0.07 per cent; tests at low vapor pressure (order of 0.2 mb) an error of slightly more than 1 per cent. In terms of dew-point measurements, the accuracy is better than 0.2°C in the range from 0 to 40°C, and better than 0.5°C in the range from −40 to 0°C.

The response time is limited by the flow rate of the air through the cavities. The time constant of the hygrometer (time to reach 63 per cent of the final value) is approximately 10 sec. In 15 sec the recorded trace reaches 90 per cent of its final value.

The wide range and the accuracy and sensitivity of this method are reached by no other system. The hygrometer is suitable as a (secondary) standard for water-vapor pressure measurements, but the technical requirements are considerable.

A microwave refractometer of somewhat different construction has also been described by C. M. Crain, *Phys. Rev.*, **74**, 691 (1948), and *Rev. Sci. Instr.*, **21**, 456 (1950); C. M. Crain and A. P. Deane, *Rev. Sci. Instr.*, **23**, 149 (1952).

## 1-73. SYSTEMS FOR HUMIDITY DETERMINATION IN LIQUIDS OR SOLIDS

*a. Dielectrical Systems.* Both the resistivity and the dielectric constant of dielectric liquids and solids vary with their water content. Increased humidity in these substances usually has the effect of decreasing their resistivity and increasing their average dielectric constant. The dielectric constant of mixtures of pure dielectric liquids and water can be computed from a simple mixing rule. The variation with humidity of the electrical properties of more complex materials, in particular those of organic origin, can only rarely be found by computation. Electrical moisture determination in such substances is usually based on empirical calibration procedures. Even so, the results of measurements may vary strongly from one specimen to another, and frequently they depend upon the concentration variation of impurities. Small amounts of dissociable salts can change the resistivity of a substance at a given moisture content by large factors.

Despite this difficulty, electrical moisture determination has found many, in particular industrial, applications for materials such as minerals, sand, coal, oil, salt, soap, food and dehydrated food products, coffee, tobacco, grain, flour, starch, wood, pulp and paper, fibrous materials, and textiles.

An example for the resistance variation with humidity in solids is shown in Fig. (1-7)13, in this case for wood. The resistance decreases nearly exponentially with increased moisture content. However, the calibration is different for different types of wood, and even for the same type of wood, for samples of different origin. The resistance of wood also varies with the direction of the current; in the direction of the grain the resistance is about $\frac{1}{10}$ of that in the direction perpendicular to the grain.

All electrical methods of moisture determination are strongly influenced by temperature variation. In wood, an increase of temperature from 0 to 30°C reduces the resistance by a ratio 5 to 1.

The resistivity can be measured with d-c methods, the electrodes being in direct contact with the test specimen, Fig. (1-7)14a. Large voltages (100 volts) are frequently used, and polarization may be negligible. In some cases difficulties arise in obtaining a reliable contact between the measuring electrodes and the specimen, in particular if the specimen has a granular structure. It is then more advantageous to measure the resistivity of the sample in a capacitor, as in Fig. (1-7)14b, by a-c methods at audio- or radiofrequencies. The equivalent circuit of such a setup is shown in Fig. (1-7)14d; the variation with

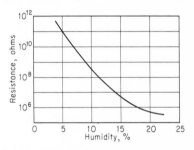

FIG. (1-7)13. Resistance of wood between two electrodes plotted versus the relative humidity (*from J. Stanek, Arch. tech. Messen*, V 1281–5, (*November*, 1935; *by permission*).

moisture of the resistance $R_s$ or of the dielectric constant (capacitor $C_s$) or of their combination (storage factor, dissipation factor, loss factor, loss tangent) can be detected and measured with any bridge arrangement. The presence of the capacitor $C_F$ in the circuit in

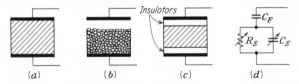

FIG. (1-7)14. Electrode arrangements for humidity determination in solids: (a) electrodes in contact with the test object; (b) electrode separated by an air space from the object; (c) insulated electrodes; (d) equivalent circuit; $C_F$, capacitance between electrode and sample; $C_s$ and $R_s$, capacitance and resistance of the sample.

series with the specimen proper does not influence the result. Figure (1-7)14d represents also the equivalent circuit for the frequently used capacitor arrangement with insulated electrodes, Fig. (1-7)14c, which avoids a direct contact between the sample and the test electrodes.

The general behavior of complex dielectric materials as a function of the applied frequency is shown in Fig. (1-7)15. The dielectric constant decreases with increased frequency, while the loss (conductive component) increases.[1] The location of the point of inflection depends upon the sample material. Increase of the temperature of the sample has the effect of shifting the point of inflection toward higher frequencies. The inflection can take place in a narrow frequency region, but in most cases it extends over a frequency range of several

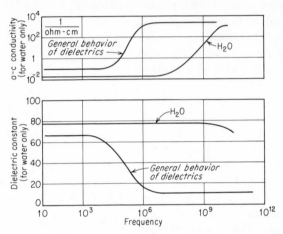

FIG. (1-7)15. Variations of conductivity and dielectric constant of dielectrics (generalized curve) and of water.

decades. The characteristic curves for water are shown in the same diagram. It can be seen that under certain circumstances measurements at two different frequencies can lead to a better definition in moisture determination.

*b. Nuclear Magnetic Resonance Method.* If hydrogen nuclei are brought into a magnetic field, they will be oriented in two different directions, depending upon their angular momentum. The two orientations correspond to two slightly different energy levels; therefore, a transition from the lower to the higher energy level will require the absorption of energy. The energy can be supplied from an electromagnetic field. The frequency at which energy is absorbed by proton nuclei is $f_0 = 4,258 \times B$, where $B$ is the flux density of the magnetic field. For practical magnetic fields of the order of several thousand gauss, the frequency is in the megacycle range. The energy absorption is measured by an electrical system (3-4); the absorption curve has the form of a narrow resonance curve ("absorption line").

[1] P. O. Schupp, *Wiss. Veröffentl. Siemens-Werke*, **17**, 1 (1938).

The intensity of the absorption line, or the height and slope of the resonance curve, depends upon the number of nuclei present in a sample. This effect forms the basis of a quantitative determination of water, i.e., proton nuclei in organic substances.[1] The sample is placed in a container inside a coil which is supplied by rf current. Coil and sample are in highly homogeneous magnetic field, Fig. (1-7)16. The variation of the rf energy absorption, in response to a

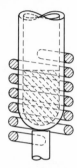

FIG. (1-7)16. Coil and sample of an arrangement for the measurement of humidity from nuclear magnetic resonance [*from J. M. Shaw and R. H. Elsken, J. Appl. Physics,* **26,** 313 (1955); *by permission*].

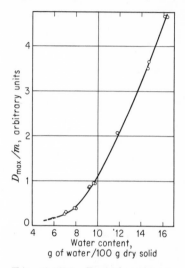

FIG. (1-7)17. Typical moisture-calibration curve for dehydrated potatoes [*from J. M. Shaw and R. H. Elsken, J. Appl. Physics,* **26,** 313 (1955); *by permission*].

variation of the frequency of the rf source or variation of the magnetic flux density, is measured by an rf bridge system.

The method has been used for organic matter (dehydrated vegetable tissue) in a humidity range (mass of water to mass of dry tissue) varying from 1 to 5, for wood having between zero and 1 gram of water per gram of dry material, for dehydrated potato powder, wheat kernels, and starch. A typical calibration curve is shown in Fig. (1-7)17.

As in other applications of the nuclear-resonance method, a high degree of magnetic-field homogeneity is required. Inhomogeneity can cause considerable error, more for granular substances than for homogeneous liquid or gaseous substances, because different grains

[1] T. M. Shaw and R. H. Elsken, *J. Chem. Phys.,* **18,** 1113 (1950).

are exposed to different field strengths. This effect causes a variation of the result, depending upon the mass of material contained in the sample. The problem is discussed in detail by Shaw and Elsken.[1] With suitable magnetic fields, the maximum error in the determination of humidity can be kept as low as $\pm 0.3$ per cent. The requirements for instrumentation and accessory equipment for this method are high.

[1] T. M. Shaw and R. H. Elsken, *J. Appl. Physics*, **26**, 313 (1955).

# 2

# *Temperature Transducers*

The two most widely used temperature transducers are the resistance thermometer and the thermocouple or thermoelement. Of these, the resistance thermometer (2-11) has potentially the highest sensitivity; it permits the detection of temperature variations in the order of 1/10,000 of one degree centigrade, at room temperature. One form of the resistive temperature transducer, the platinum resistance thermometer, forms the basis of the International Temperature Scale from $-190$ to $+660°C$. Metallic resistance thermometers (2-11a) can be used in the range from approximately $-270$ to more than $1000°C$. Semiconducting resistance thermometers (thermistors, 2-11b) are generally used in a range from $-100$ to $+300°C$, although special materials have been employed for very low temperatures from 0.15 to $20°K$ (carbon) and also at very high temperatures between 800 and $1100°C$. The sensitivity (change of resistance per unit change of temperature) of thermistors is usually considerably higher than that of metals, but their stability and accuracy are generally less than those of metallic resistance thermometers. The electrolytic transducer (2-11c) is of limited practical importance but may be useful for some biological applications. The application of the ionized-gas transducer (2-11d) is limited to the range of very high temperatures, exceeding $4000°C$.

Two forms of inductive temperature transducers have been used (2-12): The first (2-12a), based on susceptibility variations of paramagnetic salts, is applicable in the range below $4°K$; the other, (2-12b), based on susceptibility variations of ferromagnetic materials, has apparently found only one practical application in the range between room temperature and the curie point, but seems potentially a useful method for special purposes.

Also the capacitive temperature transducer (2-13), which is based upon the variation with temperature of the dielectric constant of different materials, has found limited practical applications. However, considering the high sensitivity with which capacitance variations can be measured, it is potentially a useful system.

Thermoelectric temperature transducers are described in 2-14. The range of application of these transducers extends from about $-200$ to about $+2400°C$, i.e., far into the range of optical pyrometry. The platinum–platinum rhodium thermoelement is used for the definition of the International Temperature Scale between 630.5 (zinc point) and 1063°C (gold point).

The noise thermometer (2-15) represents a promising method of absolute electric temperature measurement over a wide temperature range (100 to over 1500°K). However, the experimental difficulties connected with this method at the present are considerable.

There are, of course, indirect electrical methods of temperature measurements, such as the combination of a bimetallic strip with a displacement transducer. Such methods are not discussed in this book.

For general information on thermal transducers, see American Institute of Physics, "Temperature, Its Measurement and Control in Science and Industry," Reinhold Publishing Corporation, New York, 1941.

## 2-11. RESISTIVE TEMPERATURE TRANSDUCERS (RESISTANCE THERMOMETER)

A resistance thermometer consists of a resistive element exposed to the temperature to be measured; the resistance varies with temperature, and the resistance variation is detected in subsequent stages. In general the resistivity of metals increases with increased temperature (positive resistance-temperature coefficient); the resistivity of electrolytes, semiconductors, and insulators decreases with increased temperature (negative coefficient). Within narrow ranges of temperature where the resistance-temperature coefficient may be considered constant, the resistance of a conductor at the temperature $t$ is

$$R_t = R_0[1 + \alpha(t - t_0)] = R_0(1 + \alpha \, \Delta t) \tag{1}$$

where $R_0$ is the resistance of the conductor at the temperature $t_0$, $\alpha$ the temperature coefficient at $t_0$, and $\Delta t = t - t_0$. For larger temperature ranges the resistance follows more accurately the form

$$R_t = R_0(1 + \alpha \, \Delta t + \beta \, \Delta t^2) \tag{2}$$

Table 7 shows values of the resistance-temperature coefficient $\alpha$ for different materials at room temperature. Values of $\alpha$ reported in the literature vary considerably.

TABLE 7.	RESISTANCE-TEMPERATURE COEFFICIENTS $\alpha$
AT ROOM TEMPERATURE, $°C^{-1}$

| | | | |
|---|---|---|---|
| Nickel | 0.0067 | Gold | 0.004 |
| Iron (alloy) | 0.002 to 0.006 | Platinum | 0.00392 |
| Tungsten | 0.0048 | Mercury | 0.00099 |
| Aluminum | 0.0045 | Manganin | $\pm 0.00002$ |
| Copper | 0.0043 | Carbon | $-0.0007$ |
| Lead | 0.0042 | Electrolytes | $-0.02$ to $-0.09$ |
| Silver | 0.0041 | Semiconductor | $-0.068$ to $+0.14$ |
| | | (thermistors) | |

The resistance-temperature coefficient $\alpha$ can be found from two measurements of temperature and corresponding resistance: $R_1$ at the temperature $t_1$ and $R_2$ at the temperature $t_2$. Under the assumption that the resistance follows the linear equation (1), the coefficient is

$$\alpha = \frac{R_2 - R_1}{R_1 t_2 - R_2 t_1}$$

The resistance variation is usually measured with bridge arrangements, such as the Wheatstone bridge or the Thompson (Kelvin) bridge. A number of special bridge circuits which eliminate the influence of the supply voltage, of the temperature error of the indicating meter, or of the line connecting the resistance element with the reading setup are described by Eggers.[1] For bridges and circuits for the measurements of temperature differences with two resistance thermometers, see the paper by Eggers and also the paper by Geyger.[2] The use of the ratio meter for resistance-temperature measurements is described by Lorenz.[3] The potentiometer method for resistance measurement furnishes very accurate values over a large range of resistance variation, but is in general less convenient. The use of resistance-thermometer elements with separate voltage $E$ and current $I$ terminals as shown in Fig. (2-1)1 is recommended for potentiometric measurements to eliminate the influence of a variable resistance at the contacts.

[1] H. R. Eggers, *Arch. tech. Messen*, J 222–1, March, 1941.
[2] W. Geyger, *Arch. tech. Messen*, V 2166–2, 1931.
[3] J. Lorenz, *Arch. tech. Messen*, V 212–1, November, 1939.

Through the use of fixed series and parallel combinations of resistances, the characteristics of resistance thermometers can be altered to approximate a prescribed function, e.g., a linear one.[1]

*a. Metallic-resistance Temperature Transducers.* Most metals show an increase of resistivity with temperature which is first linear and then increases in accelerated fashion. In particular iron,

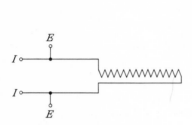

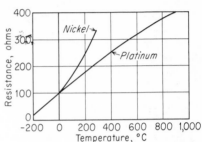

FIG. (2-1)1. Resistance thermometer, schematic diagram: *I,I*, current terminals; *E,E*, voltage terminals.

FIG. (2-1)2. Resistance variation with temperature of a platinum and a nickel resistance thermometer having a resistance of 100 ohms at 0°C.

nickel, and cobalt behave in this way. Only platinum shows a less rapid increase in resistivity at higher temperatures, as shown in Fig. (2-1)2.

The resistance-temperature coefficient of all metals depends to a large extent upon their purity and the thermal treatment. Pure metals have relatively high coefficients; the coefficients of alloys are usually smaller and can even be negative in certain ranges of temperature (manganin). Change of the physical character of the metal (annealing, recrystallization) can cause a change of the temperature coefficient which in some cases, e.g., at the curie point, can be discontinuous.

Nickel can be used below 300°C. Tungsten and tantalum have high temperature coefficients and high melting points but oxidize in air at temperatures above 400°C. The resistance variation of commercial copper can be used to determine the temperature of machine or transformer windings. In the range below 120°C a gold–silver alloy can be used which has the same characteristics as platinum.[2] For low-temperature measurement certain phosphor-bronze alloys are satisfactory. These materials exhibit a large change of resistivity below 7°K and a small variation above this range. The sensitivity in

[1] F. Lieneweg, *Arch. tech. Messen*, J 222–2, December, 1949.
[2] G. Keinath, *Arch. tech. Messen*, J 221–1, January, 1933.

the usable range is about 55 times that of an ordinary resistance thermometer. The effect seems to be caused by small amounts of lead impurity in the wire, which gradually become superconducting.[1] At low temperature the resistance is affected by the presence of magnetic fields. Varying amounts of lead in the wires allow their use in intermediate or strong fields.[2]

Several forms of resistance thermometers have been developed for the measurement of surface temperatures. The resistance wire is mounted in grid form on the flat end surface of an insulating cylinder; the diameter of the probe can be made as small as 1 mm diameter.[3]

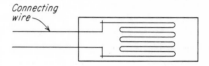

Other constructions consist of a grid of nickel wire bonded between two pieces of a flexible insulating material, as shown in Fig. (2-1)3. The resulting paper-thin sandwich is fastened by an appropriate cement, like a strain gauge, to the surface to be measured.

FIG. (2-1)3. Resistance thermometer for the measurement of surface temperature.

Surface thermometers have also been made in the form of ribbons or blankets[4] or in the form of "spaghetti."[5] They are applied to the surface under investigation by wrapping or by an adhesive tape. The lower limit of temperature for which surface thermometers can be used is about −200°C, the upper limit +250°C, although surface thermometers of this type that can be operated in the vicinity of 500°C have been described.

The platinum resistance-thermometer element usually consists of a platinum wire wound in the form of a free spiral or held in place by an insulating carrier. It is frequently enclosed in a tube for protection from mechanical injuries and chemical alteration. Of all metals, platinum is most appropriate for a resistance thermometer, because it is commercially available in pure form and is relatively stable under different environmental conditions. The relation between its resistance and temperature is simple and holds over a wide temperature range with high precision. The platinum resistance thermometer forms the basis of the International Temperature Scale from −190 to +660°C, but can be used for measurements up to 1000 and down to −264°C (9°K) when corrections are applied.

[1] J. D. Babbit and K. Mendelssohn, *Phil. Mag.*, **20**, 1025 (1935).

[2] H. Van Dijk and W. H. Keesom, *Physica*, **7**, 970 (1940).

[3] A. C. Ruge Associates, Inc., Cambridge, Mass.

[4] Minco Products Inc., Minneapolis, Minn.

[5] Trans-Sonics, Inc., Burlington, Mass.

The relation between resistance and temperature above $0°$ is given by the Callendar equation.[1]

$$t = \frac{1}{\alpha}\left(\frac{R_t}{R_0} - 1\right) + \delta\left(\frac{t}{100} - 1\right)\frac{t}{100} \qquad (3a)$$

or

$$t = \frac{R_t - R_0}{R_{100} - R_0} \times 100 + \delta\left(\frac{t}{100} - 1\right)\frac{t}{100} \qquad (3b)$$

where $t$ is the temperature in degrees centigrade, $R_t$, $R_0$, and $R_{100}$ are the resistance values at the temperatures $t°$, $0°$, and $100°C$, respectively. For very pure platinum, $\alpha = 0.00392$ and $\delta$ is between 1.49 and 1.50. For temperatures below 0 down to $-190°C$, a further correction is furnished by Van Dusen.[2] A discussion of the accuracy of the Callendar and Van Dusen equations is furnished by Mueller.[3]

The resistance thermometer is not used as a standard below $-190°C$ because in this range the resistance-temperature characteristic is greatly affected by impurities and each resistor requires a separate calibration. The National Bureau of Standards maintains a group of platinum resistance thermometers calibrated in the range from 11 to $90°K$. Other thermometers may be calibrated against these standards.[4]

TECHNICAL DETAILS. The diameter of the wire varies from 0.02 to 0.6 mm, the latter for temperatures of $1000°C$; for general use, the preferred type of wire is about 0.1 mm in diameter. For thermometers that closely duplicate those of the National Bureau of Standards, the wire should be smooth, free of defects, and drawn from an ingot which has been completely fused (not from a forged sponge). It should conform to the purity test, that is, the ratio of the resistance at $100°C$ to the resistance at $0°C$ should be larger than 1.391. The character of the metal may be tested by melting the end of the wire in a flame; it should melt quietly, without sputtering, scintillating, or evolving volatile material. In general, for stability the wire should be heated to incandescence for some time (E. F. Mueller, *loc. cit.*). The wire is then formed into a coil and mounted on a framework such as a mica cross. For methods of mounting, see J. A. Beattie et al., *Proc. Am. Acad. Arts Sci.*, **66**, 167 (1930–1931); also C. H. Meyers, *J. Research Natl. Bur. Standards*, **9**, 807 (1932). For calorimetric work, the form described by T. S. Sligh, *Natl. Bur. Standards Sci. Papers*, **17**, 49 (1922), is commonly used. A very compact form of a resistance thermometer is described by Meyers.

[1] H. L. Callendar, *Phil. Trans. Roy. Soc. (London)*, **178**, 160 (1887).

[2] M. S. Van Dusen, *J. Am. Chem. Soc.*, **47**, 327 (1925).

[3] E. F. Mueller, in "Temperature," pp. 162ff., Reinhold Publishing Corporation, New York, 1941.

[4] H. J. Hoge, in "Temperature," p. 141, Reinhold Publishing Corporation, New York, 1941; H. J. Hoge and F. G. Brickwedde, *J. Research Natl. Bur. Standards*, **22**, 351 (1939).

The total resistance of the element can vary from 0.1 to hundreds of ohms. Elements of thin wire and high resistance are preferred for low-temperature measurements, since otherwise the resistance variation is too small for convenient measurement. The thermometer element should be aged at the highest or lowest temperature to which the thermometer is to be exposed, until the resistance at 0°C remains unaltered.

In constructing a resistance thermometer, care should be taken to exclude excessive water vapor, since it is likely to be a cause of leakage resistance. Some workers recommend baking the support to

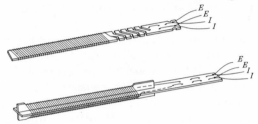

FIG. (2-1)4. Construction of resistance thermometers.

eliminate moisture. Annealing of the platinum wire and of the leads after the wire has been applied to the carrier is recommended in order to prevent strain. Protective tubes for resistance thermometers can be made from glass, quartz, porcelain, or metal (nickel). For high-temperature work, the tube is usually filled with air at $\frac{1}{3}$ to $\frac{1}{2}$ atm; for low temperature, the inside pressure should exceed the outside pressure by about 1 atm. The protecting tube at low temperature should be gas-tight to avoid condensation of water. For use below $-190°C$ the tube should be filled with helium to avoid condensation of air. For the connecting wire inside the protecting tube, gold or silver is satisfactory. The external leads are usually made from copper and connected to copper terminals, which should be mounted close together in order to reduce the influence of thermoelectric effects. Figure (2-1)4 shows some practical constructions of resistance thermometers.

Resistance thermometers are the most accurate of all temperature-sensing devices. In the vicinity of 0°C or at room temperature, measurements within an accuracy of $10^{-4}°C$ can be made. Ordinarily, however, the error is several thousandths of a degree. At 450° the accuracy does not exceed several hundredths of a degree, and around 1000° the accuracy is not higher than about one-tenth of a degree. To obtain an accuracy of 0.001°C, except at low temperatures, a precision of resistance measurements of 2 to 4 parts in a million is required.

The resistance thermometer is primarily applicable for the measurement of small temperature differences where the output from a thermocouple would be too low for convenient measurement, but it is also very accurate for the measurement of large temperature differences and high temperatures. Errors resulting from undesired thermoelectric potentials at the connections may be detected by reversing the polarity of the supply voltage; a deflection of the null detector, when the polarity is reversed, indicates a thermoelectric effect. The error can usually be corrected by taking the average of both readings.

Disadvantages of resistance thermometers are the need for auxiliary apparatus and power supplies, and the relatively large size which makes it impossible to measure temperature in a small volume, as with a thermocouple or a small thermistor.

If the measuring current is too high, it will heat the transducer element and cause an error. This condition is indicated when the temperature reading changes with a change of current through the transducer. If a measurement with the current $I_1$ furnishes a reading of temperature $t_1$ and a measurement with $I_2$ a reading of $t_2$, the corrected temperature[1] is given by

$$t = t_1 + (t_2 - t_1) \frac{I_1^2}{I_1^2 - I_2^2}$$

A current of less than 10 mA through the transducer is usually safe if the resistance wire is thicker than 0.05 mm and is immersed in a medium of high thermal conductivity (water).

*b. Semiconducting-resistance Temperature Transducers (Thermistors).* Thermistors are solid semiconductors, usually with a large negative resistance-temperature coefficient in the order of −5 per cent change of resistance per degree centigrade of temperature change (i.e., about 10 times as large as that of metals). Also thermistors with a positive temperature coefficient as high as +14 per cent/ °C in a range from 50 to 225°C have been made.[2] Thermistors consist of mixtures, mainly of the oxides but also of other compounds (sulfides, silicates) of Mn, Ni, Co, Cu, U, Fe, Zn, Ti, Al, or Mg. Carbon is also used. Thermistors are in general produced by compressing mixtures of such compounds in powder form to beads, rods, or disks

[1] H. Moser, in F. Kohlrausch, "Praktische Physik," 20th ed., sec. 4.162, 3. B. G. Teubner, Verlagsgesellschaft m.b.H., Stuttgart, 1955.

[2] H. A. Sauer and S. S. Flaschen, "Proceedings IRE-RETMA-AIEE-WCEMA Electronics Components Symposium," pp. 41–46, Engineering Publishers, Babylon, N.Y., 1956.

and sintering the mixture at high temperature into a solid mass. Electrodes are applied by firing on metal colloids, by applying metal pastes or paints, or by shrinking the material around wires. Typical forms of thermistors are shown in Fig. (2-1)5. Depending upon the composition, the resistivity of thermistor materials can have values ranging from $10^{-1}$ to $10^9$ ohm-cm, but it is usually in the range

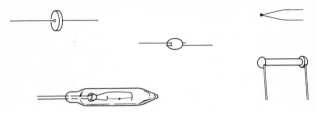

FIG. (2-1)5. Typical thermistors.

between $10^2$ and $10^6$ ohm-cm. Small admixtures of impurities or physical alterations caused by heat treatment can result in a permanent change of the resistivity by large factors. The resistivity decreases, in general, with the amount of impurities present. Figure (2-1)6 shows the variation of the resistivity with temperature of two different thermistor materials, and also, for comparison, of platinum.

Between 0 and 300°C the resistance of the thermistors decreases by a factor of about 1,000, while the resistance of platinum increases in the same range by a factor of 2.

The resistivity variation follows approximately the empirical equation

$$\rho_t = \rho_0 e^{(B/t - B/t_0)}$$

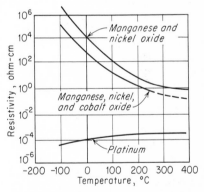

FIG. (2-1)6. Resistivity of platinum and of two thermistor materials versus temperature.

where $\rho_t$ is the resistivity at the temperature $t$, and $\rho_0$ is the resistivity at the temperature $t_0$ ($t$ and $t_0$ are expressed in degrees Kelvin). $B$ is a constant which varies for different materials and is usually in the order of 4,000. A more precise expression for the resistivity is given by Becker, Green, and Pearson.[1]

[1] J. A. Becker, C. B. Green, and G. L. Pearson, *Trans. AIEE*, **65**, 3 (November, 1949); see also O. J. M. Smith, *Rev. Sci. Instr.*, **21**, 344 (1950).

Table 8 (taken from the same publication) shows values of the resistance $R$, the temperature coefficient $\alpha$, and the constant $B$ for a typical rod-shaped thermistor element.

TABLE 8.  CHARACTERISTIC VALUES OF A THERMISTOR AT DIFFERENT TEMPERATURES

| Temp. $(t_0)$, °C | Resistance, ohms | $\alpha$, °C$^{-1}$ | $B$, °K |
|---|---|---|---|
| −25 | 580,000 | −0.061 | 3780 |
| 0 | 145,000 | −0.052 | 3850 |
| 25 | 46,000 | −0.044 | 3920 |
| 50 | 16,400 | −0.038 | 3980 |
| 75 | 6,700 | −0.033 | 4050 |
| 100 | 3,200 | −0.030 | 4120 |
| 150 | 830 | −0.024 | 4260 |
| 200 | 305 | −0.020 | 4410 |
| 275 | 100 | −0.015 | 4600 |

The voltage-current characteristics of thermistors are linear up to the range where the current causes a significant increase of the thermistor temperature. For higher currents, the characteristics become nonlinear, as shown in Fig. (2-1)7. The point at which the voltage-current characteristics start deviating from linearity depends upon the dimensions of the thermistor and the environmental conditions (thermal contact). The power dissipated before the resistance decreases by as much as 1 per cent is about 1 to 10 mW/cm$^2$ of surface in still air.

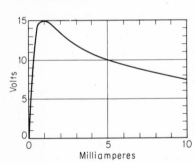

FIG. (2-1)7.  Voltage–current characteristic of a thermistor.

The time constant of a thermistor (i.e., the time in which a thermistor reaches 67 per cent of its steady-state resistance value after being exposed to a new temperature) varies for commercial thermistors between 1 and 200 sec, although occasionally figures in the millisecond range are found. For a study on the dynamic characteristics of thermistors, see Smith.[1]

The range of temperatures in which thermistors have been used extends from −100 to +300°C, although for high-stability requirements an upper limit of 100 to 150°C is frequently recommended.

[1] Smith, *ibid.*, p. 351.

Gruess[1] has used thermistors made of sintered aluminum oxide at temperatures between 800 and 1100°C. The transducers have a lifetime of several months.

Carbon resistors have been used for the measurement of very low temperatures (below 20°K) either in the form of carbon strips consisting of a thin conductive layer of carbon granules held with a binder and mounted on the object to be measured[2] or in the form of commercial radio resistors of about 2 to 150 ohms.[3] Both types exhibit a large increase of resistance in the range from 20 down to almost 0.15°K. The useful range of the carbon-strip resistor may be varied by varying the size of the carbon granules. The radio resistor is, in general, more stable than the carbon strip; in the range from 20 to 1°K the resistance value is reproducible within about 0.2 per cent. Both types are relatively insensitive to the influence of magnetic fields. They should be sealed off from exposure to liquid helium.

Most thermistors of the oxide type exhibit an aging effect, i.e., an increase of resistance with time. About 0.5 to 1.5 per cent of the resistance variation takes place during the first day or week of its use, but changes of as much as 1 per cent may still occur after several months. Preaging (exposing the thermistor to a temperature slightly higher than that to which it is to be subjected) is recommended where high stability is required; preaged thermistors may vary by as little as 0.2 per cent/year. Thermistors with pure electronic conduction exhibit less variation with time than such with ionic or mixed conduction. Enclosure of the thermistor in glass or other suitable materials to exclude chemical alterations diminishes the aging effect.

The advantages of thermistors over metal resistance thermometers are, besides their high temperature coefficients, their relatively small size, which makes localized temperature measurements possible. Thermistor beads ranging in diameter from 0.006 to 0.1 in. are commercially available. Because of their low specific heat, thermistors draw practically no heat from the object to be measured. Temperature differentials as low as 0.001°F have been measured with thermistors. Besides this, thermistors are physically rugged and have large resistance values which permit adequate impedance matching with the associated electronic equipment and, for many

---

[1] H. Gruess, *Arch. tech. Messen*, J 221–2, December, 1937.

[2] W. F. Giauque, J. W. Stout, and C. W. Clark, *J. Am. Chem. Soc.*, **60**, 1053 (1938); T. H. Geballe et al., *Rev. Sci. Instr.*, **23**, 489 (1952).

[3] J. R. Clement and E. H. Quinnel, *Rev. Sci. Instr.*, **23**, 213 (1952), and J. R. Clement et al., *Rev. Sci. Instr.*, **24**, 545 (1953).

practical purposes, eliminate the influence of the lead-wire resistance upon the temperature measurements.

For literature references, see Smith, *op. cit.*, p. 350, and H. Kanter, *Arch. tech. Messen*, Z 119–3, July, 1955; Z 119–4, August, 1955; Z 119–5, September, 1955; Z 119–6, December, 1956.

*c. Electrolytic-resistance Temperature Transducers.* The electric resistance of electrolytes decreases with increased temperature. The resistance variation is, in general, of the order of $-2$ per cent/°C but

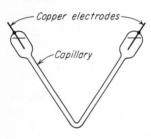

FIG. (2-1)8. Electrolytic temperature transducer.

can reach values as high as 8.9 per cent/°C (for a 42.7 per cent sodium hydroxide solution). Polarization, the development of gas, and instability are likely to cause difficulties. The use of a-c circuits is recommended in order to reduce polarization.

A stable electrolytic sensing element that can be used for d-c operation is described by Craig.[1] It consists of a solution of cuprous chloride, hydrochloric acid, and ethyl alcohol in a capillary tube with 1 mm bore. The electrodes are made of copper. The resistance-temperature coefficient is in the order of $-2$ per cent/°C. Temperature measurements obtained with this transducer are accurate to within 1°C. Figure (2-1)8

TABLE 9. RESISTIVITY AND RESISTANCE-TEMPERATURE COEFFICIENT
OF ELECTROLYTES AT 25°C

| Electrolyte | Resistivity, ohm-cm | $\alpha$, °C$^{-1}$ |
|---|---|---|
| $H_2SO_4$, density 1.223 | 1.211 | $-0.0147$ |
| $MgSO_4$, density 1.190 | 17.21 | $-0.00221$ |
| NaCl, saturated at 25°C | 3.979 | $-0.0205$ |
| KCl: | | |
| 1$N$ | 8.944 | $-0.0176$ |
| 0.1$N$ | 77.63 | $-0.0190$ |
| 0.01$N$ | 707.7 | $-0.0194$ |

shows an illustration of the electrolytic temperature transducer of Craig. Table 9 gives values of resistance-temperature coefficients of some electrolytes.

*d. Ionized Gases.* The electrical conductivity of a gas depends upon the degree of ionization, which, in turn, depends upon the

[1] D. N. Craig, *J. Research Natl. Bur. Standards*, **21**, 225 (1938).

temperature of the gas. The relation between the degree of ionization and the temperature has been derived by Saha.[1] The conductivity of the gas becomes measurable in the region above about 4000°K.

The relationship between ionization and temperature has been used by Suits[2] to measure the temperature in an electric-arc discharge operating between 1 and 10 atm of pressure. The accuracy is of the order of 10 per cent. The method is inherently suitable for the measurement at very high temperatures above ten thousand degrees Kelvin.

## 2-12. Inductive Temperature Transducers

*a. Variation of Susceptibility of Paramagnetic Salts.* The magnetic susceptibility $\chi$ of certain paramagnetic materials changes with temperature according to Curie's law:

$$\chi = \frac{C}{T}$$

where $C$ is the curie constant (a characteristic of the paramagnetic material) and $T$ the absolute temperature. These changes with temperature of the susceptibility of such salts as iron ammonium alum or iron ammonium sulfate which have a very low curie point, can be used to measure extremely low temperatures in the range below 4°K.

The susceptibility is usually determined by means of an arrangement as shown in Fig. (2-1)9. A current source produces in a primary coil $P$ a magnetic field of several gauss. Two secondary windings $S_1$ and $S_2$ are wound in opposite directions and connected, as shown, to a galvanometer. In the absence of a paramagnetic specimen, the voltages induced in both secondary windings are equal and of opposite sign, and the galvanometer

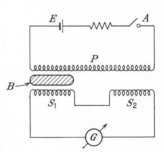

Fig. (2-1)9. System used for inductive temperature transducer, based upon susceptibility variation of a specimen $B$. $E$, voltage source; $A$, switch; $P$, primary, $S_1$ and $S_2$ secondary windings of a transformer; $G$, galvanometer.

should show no deflection when the switch $A$ is closed. If a sample $B$ of a paramagnetic salt is introduced in one coil, the voltage induced in this coil is increased, and upon closing the switch $A$ the galvanometer receives a ballistic impulse proportional to the

[1] M. N. Saha, *Phil. Mag.*, **40**, 472 (1920).

[2] C. G. Suits, *J. Appl. Physics*, **10**, 728 (1939).

susceptibility of the salt. The coils are usually kept in liquid hydrogen to keep their resistance low and constant.

The susceptibility method is the primary method for the measurement of lowest temperatures. However, the experimental requirements are considerable and the results appear in "curie temperature" and have to be corrected for the thermodynamic temperature scale.

For general information and references, see C. F. Squire, "Low Temperature Physics," pp. 147, 156–157, McGraw-Hill Book Company, Inc., New York, 1953; W. A. Allis and M. A. Herlin, "Heat and Thermodynamics," pp. 180–181, McGraw-Hill Book Company, Inc., New York, 1952.

*b. Permeability and Eddy-current Losses.* An empirical method of measuring temperatures by measuring the change of permeability and eddy-current losses in ferromagnetic materials is described by

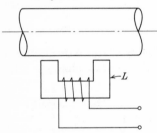

Keinath.[1] The method is primarily applicable for temperature determinations in rotating parts, such as heated steel cylinders or rollers, and does not require direct physical contact between the rotating part and the measuring device.

A schematic diagram is shown in Fig. (2-1)10. A steel cylinder of varying temperature forms the yoke of an iron-core inductor $L$ from which it is separated by a small air gap. A variation of the permeability or of the eddy-current losses resulting from a temperature change of the steel cylinder affects the magnitude of the magnetic flux and is reflected in a change of the reactive and resistive component of $L$, which can be detected in subsequent stages. Empirical calibration is required. No data on the performance of this method are available.

Fig. (2-1)10. Inductive temperature transducer.

## 2-13. Capacitive Temperature Transducer

The dielectric constant of some insulating or semiconducting materials varies with temperature; a capacitor made up with such a material as a dielectric changes capacitance with temperature and can be used as a temperature transducer.

A material of this kind, a titanate named Thermacon, is described by Dranetz, Howatt, and Crownover.[2] Depending upon its

[1] G. Keinath, *Arch. tech. Messen*, J 215–2, January, 1934.

[2] A. I. Dranetz, G. N. Howatt, and J. W. Crownover, *Tele-Tech*, June, 1949, p. 36.

composition, it can have a negative or positive temperature coefficient. Three of the specimens presented in this paper have the following properties:

*a.* Dielectric constant at room temperature, 46; temperature coefficient, −0.0003/°C.

*b.* Dielectric constant at room temperature, 490; temperature coefficient, −0.003/°C.

*c.* Dielectric constant at room temperature, 18; temperature coefficient, +0.0001/°C.

In the temperature range from −40 to +160°C, the dielectric constant changes linearly with the temperature.

Some materials (e.g., some barium titanates in the vicinity of the curie point) exhibit very large changes of dielectric constant with temperature, but show a large hysteresis of the dielectric constant versus the temperature function which limits their application to temperature-sensing elements. Other materials, such as polyvinyl alcohol, show a variation of the dielectric constant from 8 to 500 in a temperature range from 25 to 85°C as measured at 60 cps. However, these materials also show an increase of dielectric losses with temperatures of such magnitude as to make their application for temperature measurements impractical.

For numerical values of the temperature constant of some dielectrical materials, see Dwight E. Gray (ed.), "American Institute of Physics Handbook," sec. 5d, McGraw-Hill Book Company, Inc., New York, 1957.

## 2-14. Thermoelectric Transducers (Thermoelements or Thermocouples)

The contact potential between two dissimilar metals varies with the temperature of their junction. If two dissimilar metals $A$ and $B$ are joined together as shown in Fig. (2-1)11, and if their junction points $P_1$ and $P_2$ are kept at the temperature $t_1$ and $t_2$, respectively, an electromotive force (emf) arises which causes a current in the circuit. The thermal emf is to a first approximation a linear function of the temperature difference $\Delta t = t_2 - t_1$ and is independent of the temperature and the temperature gradient of the wires themselves. The thermal emf can

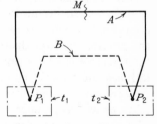

Fig. (2-1)11. Thermoelements, basic system.

be used, therefore, to measure the temperature difference $\Delta t$ between the two contact points $P_1$ and $P_2$. The order of magnitude of the thermal emf is in general several millivolts for a temperature difference of 100°C.

The emf is usually measured by cutting one of the conductors, for instance at the point $M$, and inserting a meter. Since the connecting wires are in general made from a material $C$, other than $A$ and $B$, the circuit assumes the form of Fig. (2-1)12. The sum total of all emfs in such a circuit is zero as long as all connecting points $P_1$, $P_2$, and $P_3$ are at the same temperature $t_1$. However, if the point $P_2$ is exposed to a temperature $t_2 = t_1 - \Delta t$, an emf will arise in the circuit, which is again to a first approximation proportional to $\Delta t$. The magnitude of the thermal emf is independent of the presence of the third metal $C$ as long as the contact points $P_1$ and $P_3$ are at the same temperature. If the temperature of $P_1$ and $P_3$ is not the same, the total emf developed in the circuit is equal to the algebraic sum of the thermal emfs developed at each junction.

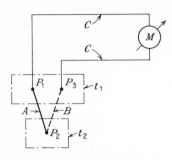

FIG. (2-1)12. Thermoelement made from two wires $A$ and $B$; junction point $P_2$ at temperature $t_2$; the wires $A$ and $B$ are connected, with a wire $C$ made from a third material, to the meter $M$. The junctions $P_1$ and $P_3$ are kept at the temperature $t_1$.

The thermal emf is usually measured either with a galvanometer (millivolt-meter) or a potentiometer, both of which can be directly calibrated in terms of temperature. The galvanometer measures the emf indirectly by measuring the current produced by the emf. Since the current depends also upon the resistance in the circuit, the measurement is correct only if the resistance is constant. Variation of temperature of the wires tends to vary the resistance, thus introducing an error in the measurement. The error can be minimized if the resistance in the circuit is large compared to such resistance variations. Since it is desirable that the greatest voltage drop is across the meter, meters with relatively high resistances are employed, although the use of such meters reduces the sensitivity of the measuring system.

A current in the thermocouple circuit tends to develop heat in the wires as well as in both junctions (Joule effect). A further influence of a current in the circuit is a temperature increase of one junction and a temperature decrease of the other junction (Peltier effect). The magnitude of these effects is frequently negligible. The current errors can be completely avoided by measuring the thermal emf with a potentiometer method.

The output voltage of the thermoelement is proportional to the temperature difference $\Delta t$ between the two junctions only within a narrow range of temperature. The output voltage $E_o$, as a function of the temperature, follows more accurately the expression

$$E_o = At + \tfrac{1}{2}Bt^2 + \tfrac{1}{3}Ct^3 \tag{1}$$

where $t$ is the temperature of one junction, the other junction being kept at a temperature of 0°C. The constants $A$, $B$, and $C$ depend upon the materials of which the thermoelement is made; for values of $A$, $B$, and $C$ see the International Critical Tables. Also Eq. (1) can be considered only as an approximation which is correct within a limited temperature range (for instance, for the platinum–platinum 90-rhodium 10 element only between 630.5 and 1063°C). All thermal emfs are reduced to zero in the vicinity of the absolute zero point.

The sensitivity (also frequently referred to as thermoelectric power $Q$) of a thermoelement is

$$S = \frac{dE}{dt} = A + Bt + Ct^2$$

and is usually expressed in microvolts per degree centigrade. The magnitude of $S$ depends upon the chemical composition and the

TABLE 10. THERMOELECTRIC SENSITIVITY $S = dE/dt$ OF THERMOELEMENT
MADE OF MATERIALS LISTED AGAINST PLATINUM, $\mu V/°C$
(Reference junction kept at a temperature of 0°C)

| | | | |
|---|---|---|---|
| Bismuth | −72 | Silver | 6.5 |
| Constantan | −35 | Copper | 6.5 |
| Nickel | −15 | Gold | 6.5 |
| Potassium | −9 | Tungsten | 7.5 |
| Sodium | −2 | Cadmium | 7.5 |
| Platinum | 0 | Iron | 18.5 |
| Mercury | 0.6 | Nichrome | 25 |
| Carbon | 3 | Antimony | 47 |
| Aluminum | 3.5 | Germanium | 300 |
| Lead | 4 | Silicon | 440 |
| Tantalum | 4.5 | Tellurium | 500 |
| Rhodium | 6 | Selenium | 900 |

physical treatment of the materials used and varies with temperature. Values given in standard tables vary considerably  Approximate values of $S$ for thermoelements made from different materials against platinum in the vicinity of 0°C are given in Table 10. The output voltage from a thermoelement made from two metals listed in Table 10 can be computed from the difference of the values of $S$. EXAMPLE: Copper versus platinum, 6.5; constantan versus platinum, $-35$; copper versus constantan, $6.5 - (-35) = 41.5 \mu V/°C$.

Because of the variation of the sensitivity $S$ with temperature, the sequence of the materials in Table 10 varies with temperature, too. Calibration curves and tables for a number of standardized thermocouple combinations are available from the National Bureau of Standards and from the Leeds and Northrup Company in Philadelphia. For thermoelements made from nonstandard materials, empirical calibration is required.

The more common types of thermocouples are described in Table 11. For detailed information concerning the properties and specification for standardization of the metals used in these thermoelements, see the monograph on thermoelectric thermometry by Dike.[1] For general information on thermoelectric thermometry, see also Roeser.[2]

The wires from which thermoelements are to be made should be annealed at a temperature higher than that at which the thermoelement is to be used. It is important that the wire is homogeneous, since otherwise a temperature gradient along the wire will cause parasitic thermal emfs (Becquerel effect). The wire can be tested for homogeneity by connecting both ends to a galvanometer and moving a bunsen burner along the wire; the galvanometer should not show any deflection. The contamination of platinum or platinum-rhodium wires can sometimes be removed by heating the wire for a few minutes to about 1300°C to oxidize the impurities. Oxides can be removed from the surface with a molten borax bead flowing down the wire.

The wires can be joined together by soldering or welding. Platinum thermoelements are usually welded (without twisting the end) in an oxygen-hydrogen flame or in a carbon arc without the use of a flux. Copper-constantan and iron-constantan thermoelements can be twisted with moderate strain and silver-soldered. Some authors recommend the use of ammonium chloride or rosin flux. Borax or fluorspar is recommended as a flux for chromel-alumel thermoelements. Thin wires can be welded by discharging a capacitor of several microfarads, charged to a voltage of 20 to 100 volts, through the junction. Soft solder (tin-lead) and low-melting alloys, such as Woods metals, are satisfactory at lower temperatures. See F. Kerkhof, *Arch. tech. Messen*, J 2404–1, October, 1940. Thermoelements must be protected from mechanical injuries, chemical contamination, and electrical disturbances (leakage, a-c pickup, galvanic emf in electrolytes). Platinum-platinum rhodium thermocouples are particularly susceptible to damage if exposed to vapors of other metals, or to hydrogen, sulfur, and carbon monoxide. Thin wires are more susceptible to contamination than thick wires. Protection in metal or ceramic tubes is required in most

[1] P. H. Dike, "Thermoelectric Thermometry," Leeds and Northrup Company, Philadelphia, 1954.

[2] W. F. Roeser, in "Temperature," pp. 180ff, Reinhold Publishing Corporation, New York, 1941.

## TABLE 11. PROPERTIES OF THERMOCOUPLES

| Materials | Useful temp. range for continuous operation, °C | Max. temp. for short periods, °C | Sensitivity output, μV/°C | Remarks |
|---|---|---|---|---|
| Pt-Pt$_{90}$Rh$_{10}$ | 0 to +1450 | 1700 | 5 at 0°C<br>12 at 1500°C<br>0 at −138°C | Defines the international temperature scale from 630.5 to 1063°C. Not suitable at low temperatures since output below 0°C is less than 5 μV/°C, and is zero at −138°C. Accuracy at gold point (1063°C) is 0.2°C and at zinc point (419.5°C) is 0.1°C. Most stable of all thermocouples between 400 and 1600°C. |
| Copper-constantan (constantan approximately Cu$_{57}$Ni$_{43}$ plus small percentage of Mn, Fe, C) | −200 to +350 | 600 | 15 at −200°C<br>60 at +350°C<br>Useful at −262°C | Sensitivity increases almost linearly. Upper limit of 350°C imposed by oxidation of copper. Preferable to iron-constantan because of greater homogeneity of Cu than Fe. Small resistance variation with temperature. |
| Iron-constantan | −200 to +800 | 1000 | 45 at 0°C<br>57 at 750°C | Widely used in industrial measurements. Superior in reducing atmospheres. |
| Chromel P-alumel (chromel P is Ni$_{90}$Cr$_{10}$, alumel is Ni$_{95}$ plus Al, Si, Mn) | −200 to +1200 | 1350 | From 40 to 55<br>Between 250 and 1000°C | More resistant to oxidation than other base metals. Most satisfactory between 650 and 1200°C. |
| Chromel P-constantan | −200 to +800 | 1000 | 250 | |
| Graphite-silicon carbide* | | 1800 | | |
| Iridium-Ir$_{50}$Rh$_{50}$ | | 2000 | | |
| Iridium-Ir$_{90}$Rh$_{10}$ | | 2400 | | |
| Tungsten-W$_{075}$Mo$_{25}$ | | 2600 | | |
| Gold-silver | −259 to −190 | 100 | 450 | |
| Tellurium-platinum† | −200 to +350 | 400 | 40 at 0°C | |
| Copper-copnic‡ | −200 to +860 | 1000 | 60 at 0°C | |
| Iron-copnic | 0 to +660 | 1000 | 75 at 0°C | |
| Chromel-copnic | | | | |

* G. R. Fitterer, *Am. Inst. Mining Met. Engrs.*, February, 1933.
† G. Keinath, *Arch. tech. Messen*, J 241-1, January, 1933.
‡ General Electric Company.

applications, but such procedure increases the time lag of the thermoelement and is likely to cause the thermoelement temperature to be different from the temperature to be measured. Metal tubes are usually adequate for the base-metal thermocouples, which are protected from contamination by their oxides, but may be entirely unsatisfactory if furnace gases are present. Sometimes two tubes are used, e.g., a tube of fused silica inside a second tube of metal, silicon carbide, or fire clay. The primary tube of low volatility is intended to be impervious to gases at high temperatures, and the secondary tube provides resistance to mechanical and thermal shock and corrosion. All ceramic materials tend to become conductive at higher temperatures. For further reference on protection tubes, see P. H. Dike, *op. cit.*, chap. 4F.

A needle thermoelement for subcutaneous and intramuscular temperature measurement is described by Krog.[1] Thermoelements with extremely small heat capacity are frequently used for radiation measurements (see 5-14a). High-velocity thermocouples for gas-temperature measurements are described by Mullikin.[2]

REFERENCE POINT. The temperature $t_2$ of the measuring junction $P_2$ in Fig. (2-1)12 can obviously only be determined if the temperature $t_1$ of the reference junctions $P_1$ and $P_3$ is constant and known.

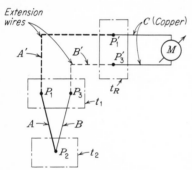

FIG. (2-1)13. Thermoelement arrangement with extension wires.

Technical difficulties in keeping the reference junctions at a constant temperature may arise, for instance, if the junction points $P_1$ and $P_3$ are in the vicinity of a furnace. It is customary, in such cases, to connect the two wires $A$ and $B$ with two extension wires $A'$ and $B'$ leading to a place $t_R$ where the reference temperature is constant, as shown in Fig. (2-1)13. The materials of $A'$ and $B'$ must be chosen so that the thermal emf at the junction points $P_1$ and $P_3$ is negligible. Such extension wires (for instance, a chromel-alumel pair) made up as an insulated duplex cable are commercially available for most practical thermocouple combinations. Because of the high price of platinum thermoelements, extension wires for platinum–platinum 90-rhodium 10 elements usually consist of copper and a nickel-copper alloy. The thermoelectric emf of these materials versus platinum and platinum rhodium is negligible within a limited temperature range.

[1] J. Krog, *Rev. Sci. Instr.*, **25**, 799 (1954).

[2] H. F. Mullikin, in "Temperature," p. 775, Reinhold Publishing Corporation, New York, 1941.

For precision work the reference junction should be kept at 0°C in a distilled-water ice bath.  In some industrial applications the reference junction is buried at a depth of 4 to 6 ft underground, where the temperature is sufficiently constant throughout the year.  For some uses it is preferable to keep the reference junction at room temperature or in a thermostatically controlled constant-temperature bath, usually at a temperature of about 50°C.

Errors resulting from variations of the reference temperature may be minimized or eliminated by manual or automatic mechanical

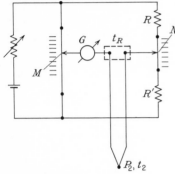
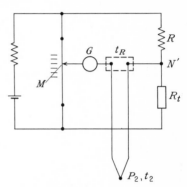

FIG. (2-1)14. Circuit for manual compensation of the error resulting from temperature variation of the reference junction $t_R$.

FIG. (2-1)15. Circuit for automatic compensation of the error resulting from temperature variation of the reference junction $t_R$.

(bimetallic-spring) adjustment of the meter or by electrical means. Figure (2-1)14 shows a potentiometer setup for manual electric compensation of the reference-temperature variations.  With the contact point $N$ in a fixed position, the thermal emf of the junction $P_2$ is read on the scale $M$ after adjusting the sliding contact $M$ until the galvanometer $G$ reads zero.  The temperature of the reference junction $t_R$ is read on a thermometer, and the influence of a temperature variation is compensated for by manual adjustment of the sliding contact $N$ on a scale $N$, usually calibrated in temperature degrees. Automatic compensation can be obtained by replacing the fixed resistors $R$ and $R'$ and the slide wire $N$ by a network containing a temperature-sensitive element $R_t$, as illustrated in Fig. (2-1)15.  The potential variation of the point $N'$ compensates for the changes in thermal emf resulting from a variation of the temperature $t_R$.  Another method of reference-point temperature compensation is shown in Fig. (2-1)16.  The three fixed resistors $R_1$, $R_2$, and $R_3$ and the temperature-sensitive resistor $R_t$ form a normally balanced bridge, which is fed from a small d-c voltage source $E_A$. The bridge becomes unbalanced for any change of the temperature-sensitive resistor $R_t$ and furnishes an output voltage which compensates for the change of thermal emf occurring at the reference junction.

A further means to reduce the effect of temperature variation of the reference junction is to use thermoelements with curved characteristics, as shown in Fig. (2-1)17. The platinum–platinum rhodium

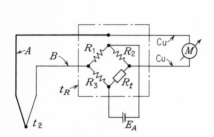

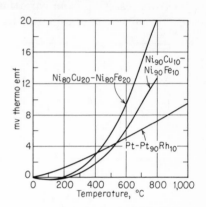

FIG. (2-1)16. Alternate circuit for reference-temperature error compensation.

FIG. (2-1)17. Thermoelements with curved characteristics which eliminate the error resulting from temperature variation of the reference junction.

thermoelement curve rises slowly in the vicinity of 0°C and faster at higher temperature, thus leading to a reduction in the error produced by the reference-temperature variation. The curves for the nickel copper–nickel iron thermoelements show almost no increase of the thermal emf between 0 and 200°C. If the reference junction of such a thermocouple is held within this range, any temperature variation will cause a reading error of less than 2°C for a reference-temperature variation of 100°C.[1]

For measuring procedures on thermocouples, see Wm. F. Roeser and T. Lonberger, Methods of Testing Thermocouples and Thermocouple Materials, *Natl. Bur. Standards Circ.* 590, February, 1958.

## 2-15. NOISE THERMOMETER

The electrons in a conductor at temperatures above the absolute zero point are in constant random motion. This motion of the

[1] W. Rohn, *Z. Metallk.*, **16**, 297 (1924); Kulbush and Kalinin, *Präzisionsindustrie* (*Moscow*), 1933, nos. 3 and 4; referred to by G. Keinath, *Arch. tech. Messen*, J 241–3, July, 1934.

electrons produces randomly varying potentials between the terminals of a conductor which increase in magnitude with the temperature of the conductor.[1] The mean square voltage $\bar{V}^2$ between the terminals of a conductor with the resistance $R$ which is held at the absolute temperature $t$ in degrees Kelvin is proportional to the product $Rt$.

Such randomly occurring potentials contain frequency components from zero to the highest measurable frequencies. If measured within a limited frequency band $\Delta f$, the mean square voltage is

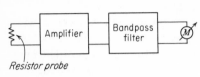

Fig. (2-1)18. Noise thermometer, schematic diagram.

$$\bar{V}^2 = 4ktR\,\Delta f$$

where $k$ is the Boltzmann constant, $1.37 \times 10^{-23}$ joule/deg. A resistor in the kilohm range at room temperature will have an rms noise voltage of a few microvolts measured in a frequency interval of $10^4$ cps.

A schematic diagram of a noise thermometer is shown in Fig. (2-1)18. In principle, a measurement of the noise voltage generated in a resistor would be sufficient to determine its temperature. However, the accurate determination of a mean square voltage in the microvolt range is technically difficult.

Garrison and Lawson[2] and Hogue[3] have successfully built noise thermometers for the 1000°K range in which they compare the unknown temperature of one resistor with the known temperature of another. Their

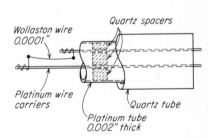

Fig. (2-1)19. Resistance probe of a noise thermometer [*from J. B. Garrison and A. W. Lawson, Rev. Sci. Instr.,* **20,** 785, (1949); *by permission*].

method is the following: A resistor $R_1$ at the known temperature $t_1$ is adjusted in value until the noise voltage from it is the same as that from another resistor $R_2$ at the unknown temperature $t_2$. When this adjustment is made, the temperature $t_2$ is given by

$$t_2 = t_1\left(\frac{R_1}{R_2}\right)$$

[1] J. B. Johnson, *Phys. Rev.,* **32,** 97 (1928).
[2] J. B. Garrison and A. W. Lawson, *Rev. Sci. Instr.,* **20,** 785 (1949).
[3] E. W. Hogue, *Natl. Bur. Standards Rept.* 3471, July, 1954.

The voltage comparison is made by repetitive switching from one resistor to the other. The output voltage is amplified, rectified, and integrated. The switching method eliminates differences in amplifier characteristics and drift. The temperature probe of Garrison and Lawson is a resistance in the kilohm range. It consists of a platinum Wollaston wire (0.0001 in.) enclosed in a protective platinum cylinder, as shown in Fig. (2-1)19, and it is suitable for temperatures up to 1400°C. For higher temperatures, tungsten can be used.

The noise thermometer can measure temperatures over a wide range (100 to 1700°K) with good accuracy; 0.1 per cent has been claimed over the entire range with very careful selection of the electronic components, while 1 per cent requires no special precautions. The resistance element itself requires no calibration. There are no thermal hysteresis effects, and any metal can be used. Carbon resistors are not suitable because of inhomogeneities in their structure. The technical difficulties in overcoming stray capacitance, pickup, and drift are considerable.

# 3

# *Magnetic Transducers*

The most widely used magnetic transducer (a transducer that responds to magnetic field strength) is the induction coil or search coil (3-11). The *stationary* search coil (3-11a) requires a variation with time of the magnetic field; it is useful whenever the field can be turned on or off or when it varies periodically with time. Where this cannot be done (e.g., the magnetic field of the earth or that in magnetic circuits containing ferromagnetic materials which show a residual flux density), the *moving* coil (3-11b) furnishes adequate results. The movement can be a translation, a rotation, or an oscillation. A particularly useful induction system is the crystal-driven oscillating loop; it is capable of measuring an area of not more than 0.15 mm². The output from the search coils is a voltage which may be constant, alternating, or a pulse. Special search coils have been devised for measurements in inhomogeneous fields (3-11c).

Transducer systems based upon the effect of a magnetic field on moving electric charges are described in 3-12. The magnetic field may act on the electrons in a moving metal disk (3-12a) or upon the charged carriers of electricity in a moving conducting liquid (3-12b), or it may act on moving free electrons (3-12c) or on electrons and positive holes moving in a crystal lattice (3-12d).

The variation with magnetization of the permeability of certain ferromagnetic materials forms the basis of a further group of transducers (3-13). The permeability variation may lead to a change of inductance (3-13a), or it may cause a variation in magnitude and direction of the magnetic flux density (3-13b); it may cause a distortion of a wave shape and, in further sequence, the generation of harmonics (3-13c), or it may lead to a phase shift of an a-c voltage (3-13d).

Another group of magnetic transducers is based upon the phenomenon of nuclear resonance (3-14).

There is a further group of transducers which do not directly furnish an electric output in response to a magnetic input, but will do so in combination with other transducers or servosystems (3-15). The direct response may be a mechanical displacement (3-15a), a temperature increase (3-15b), or it may be of an optical character (3-15c).

Magnetic fields have been measured in a range from $10^{-6}$ to about $10^6$ oersteds (flux densities $10^{-6}$ to $10^6$ gauss). The magnetic field of the earth is in the order of 0.5 oersted; its horizontal component varies between 0.15 and 0.4 oersted. Permanent magnets produce fields in the order of 100 to 10,000 oersteds. The field strengths in electrical machines and in particle accelerators are in the order of 10,000 to 20,000 oersteds. Electromagnets for laboratory use may produce fields in narrow gaps up to 60,000 oersteds. Kapitza[1] obtained field strengths in the order of 500,000 oersteds, and Wald[2] field strengths in the order of 450,000 oersteds. Furth and Wanick[3] succeeded in obtaining fields lasting from 50 to 200 $\mu$sec and reaching values as high as 600,000 oersteds. Foner and Kolm[4] have produced fields of 750,000 oersteds.

Information about the useful range of different transducer systems may be found in Table 12. The most sensitive systems are the electron-beam magnetometer and the even-harmonic method. The highest absolute accuracy is obtained with the nuclear-resonance system.

A system of a given sensitivity for use at low field strength may well be used in fields of higher intensities if the unknown field strength is compensated for by known bias field, such as the field produced by a coil of well-defined dimensions through which an adjustable current passes. The transducer system is then used as a null detector, or it measures variations of magnetic field strength.

For summarizing literature and references on magnetic fields and their measurements, see L. F. Bates, "Modern Magnetism," Cambridge University Press, London, 1948; R. M. Bozorth, "Ferromagnetism," Bell Laboratory Series, D. Van Nostrand Company, Inc., Princeton, N.J., 1951; H. Neumann, *Arch. tech. Messen,* V 391–4, August, 1940; E. C. Stoner, "Magnetism and Matter,"

[1] P. Kapitza, *Proc. Roy. Soc. (London),* (A)**115**, 658 (1927).
[2] T. F. Wald, *J. Inst. Elec. Engrs.,* **64**, 509 (1926).
[3] H. P. Furth and R. W. Wanick, *Rev. Sci. Instr.,* **27**, 195 (1956).
[4] S. Foner and H. H. Kolm, *Rev. Sci. Instr.,* **27**, 547 (1956).

Methuen & Co. Ltd., London, 1934; E. M. Underhill (ed.), "Permanent Magnet Handbook," Crucible Steel Company of America, Pittsburgh, Pa., 1957; S. R. Williams, "Magnetic Phenomena," McGraw-Hill Book Company, Inc., New York, 1931.

TABLE 12.  USEFUL RANGE OF MAGNETIC TRANSDUCER SYSTEMS

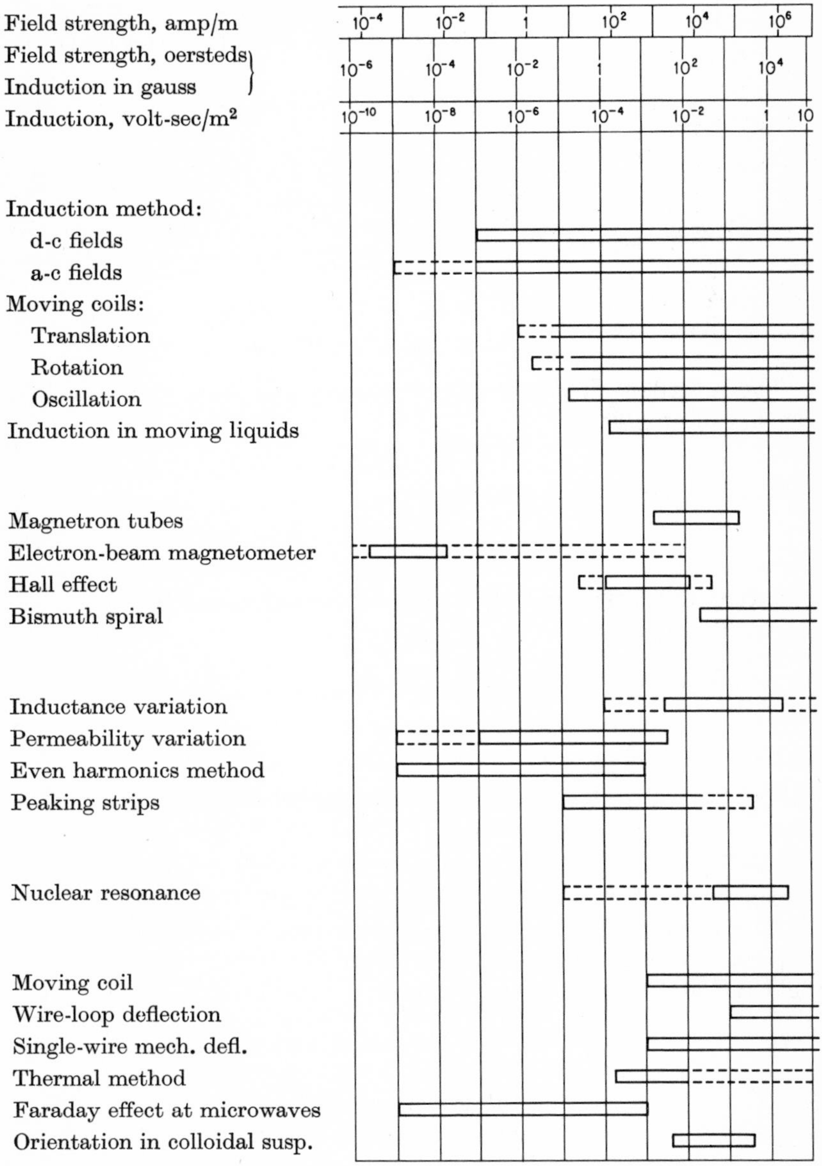

| | | | | | |
|---|---|---|---|---|---|
| Field strength, amp/m | $10^{-4}$ $10^{-2}$ 1 $10^2$ $10^4$ $10^6$ | | | | |
| Field strength, oersteds<br>Induction in gauss | $10^{-6}$ $10^{-4}$ $10^{-2}$ 1 $10^2$ $10^4$ | | | | |
| Induction, volt-sec/m$^2$ | $10^{-10}$ $10^{-8}$ $10^{-6}$ $10^{-4}$ $10^{-2}$ 1 10 | | | | |

Induction method:
  d-c fields
  a-c fields
Moving coils:
  Translation
  Rotation
  Oscillation
Induction in moving liquids

Magnetron tubes
Electron-beam magnetometer
Hall effect
Bismuth spiral

Inductance variation
Permeability variation
Even harmonics method
Peaking strips

Nuclear resonance

Moving coil
Wire-loop deflection
Single-wire mech. defl.
Thermal method
Faraday effect at microwaves
Orientation in colloidal susp.

## 3-11. INDUCTION SYSTEMS, SEARCH COILS

A coil having $n$ turns and an average area $a$ is placed in a magnetic field of uniform strength $H$, Fig. (3-1)1. The direction of the magnetic field is perpendicular to the plane formed by the loop. Any change of the magnetic field produces an instantaneous voltage $e_o$ between the loop terminals, which is proportional to the rate of change of the magnetic field

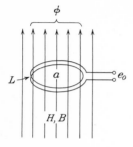

$$e_o = na\mu \frac{dH}{dt} \qquad (1a)$$

where $\mu$ is the magnetic permeability. If the direction of the magnetic field is not parallel to the coil axis but forms an angle $\alpha$ with the axis, Eq. (1a) changes to

FIG. (3-1)1. Search-coil magnetic transducer.

$$e_o = na\mu \frac{dH}{dt} \cos \alpha \qquad (2)$$

The magnetic field strength $H$ corresponds to a magnetic flux density of magnitude

$$B = \mu H \qquad (3)$$

so that Eq. (1a) can be written as

$$e_o = na \frac{dB}{dt} \qquad (1b)$$

The total flux through the area $a$ of the loop is defined as

$$\Phi = a\mu H = aB \qquad (4)$$

so that Eq. (1a) can also be written as

$$e_o = n \frac{d\Phi}{dt} \qquad (1c)$$

The Eqs. (1a), (1b), and (1c) describe the relationship between the rate of change of a magnetic flux through a loop and the voltage induced in the loop. In general, however, the input magnitudes to be converted into electrical signals are not the time derivatives of $H$, $B$, or $\Phi$ but the magnitudes themselves. The integral form of Eq. (1a) furnishes a relationship between any change of the magnetic field $\Delta H$ and the transducer output.

$$\int_{t_0}^{t_1} e_o \, dt = na\mu \, \Delta H \qquad (5a)$$

That is, the voltage-time integral appearing at the transducer output is proportional to the change of the magnetic field strength $\Delta H$

taking place during the time $t_0$ to $t_1$ and is independent of the rate of change within this period. In practice, the magnetic field strength is frequently changed from zero to the value $H$, so that $\Delta H = H$, or from the value $-H$ to $+H$, so that $\Delta H = 2H$. The integral form of Eq. (1b) gives

$$\int_{t_0}^{t_1} e_o \, dt = na \, \Delta B \tag{5b}$$

and, correspondingly, from Eq. (1c),

$$\int_{t_0}^{t_1} e_o \, dt = n \, \Delta\Phi \tag{5c}$$

The physical relationships expressed in Eqs. (1) and (5) form the basis of all induction-type magnetoelectric transducers. They differ only in the manner by which a change of flux is accomplished.

The voltage $e_o$ appearing at the output terminals can be measured with any suitable meter which measures instantaneous values. The voltage-time integral can be measured with any suitable electric integrator. Ballistic galvanometers or fluxmeters are frequently used because of the sensitivities for small voltages generated by low impedance sources. Many of the integrating methods require that the integration time $t_1 - t_0$ be short compared with the time constant of the integrating device. For this reason the rise time or decay time of the flux variation $\Delta\Phi$ must be short compared with the time constant of the integrator.

*a. Stationary Search Coils.* D-C FIELDS. If the input field can be turned on or off or can be reversed or varied between defined limits, a stationary coil can be used as a transducer and will furnish an output according to Eq. (1a) or (5a). The method is not applicable in the presence of ferromagnetic materials, which may cause a residual flux of unknown magnitude. There is no restriction in size, shape, or number of turns of the coil, except that the coil should be so small that the field is homogeneous over its area. A larger coil, while still measuring correctly the total flux through its area, will produce an output proportional to the average field strength or flux density over the area of the coil. (For an exception, see the flux ball, 3-11c).

The constant of the coil, $na$, can be found with fair accuracy from the coil dimensions. Kussmann[1] recommends the determination of $na$ by counting the number of turns $n$ and measuring the length of the (unwound) wire. Higher accuracy may be obtained

---

[1] A. Kussman, in F. Kohlrausch, "Praktische Physik," H. Ebert and E. Justi (eds.), 20th ed., p. 220, B. G. Teubner Verlagsgesellschaft m.b.H., Stuttgart, 1956.

by calibrating the coil in a magnetic field of known field strength, such as the inside of a long induction coil. An accuracy of 0.05 per cent can be obtained.

The search coil method is applicable for fields from $10^{-3}$ oersted up to the highest available field strengths. For fields of a strength less than $10^{-3}$ oersted, the output voltage is usually too small for convenient measurement.

A-C FIELDS. If the field strength varies sinusoidally with time, the induced voltage is sinusoidal too and can be measured, either directly or after amplification, with a-c instruments. The method is also applicable to fields of periodic but not sinusoidal character if the necessary corrections for wave shape are applied. The sensitivity of the a-c magnetic field strength measurement can be very high. Fields of the order of $10^{-5}$ oersted have been measured.[1] Stray magnetic fields, mostly of line frequency but also of higher harmonics, can introduce considerable error. The error can be reduced by magnetic shielding of the source of the stray field or eliminated by placing the axis of the pickup coil at an angle of 90° with respect to the stray field.[2] It is also possible, though experimentally difficult, to eliminate the influence of the stray field by compensation with a second field of proper magnitude, phase, and wave shape produced by an auxiliary coil.

For low frequency of the a-c field, the output from the coil is frequently too low for convenient detection. At high frequencies in the rf range, the capacitive coupling between adjacent turns of the pickup coil tends to reduce the output.

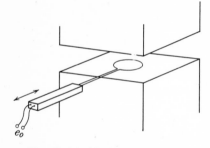

FIG. (3-1)2. Moving search coil.

*b. Moving Coils.* When the magnetic field strength cannot be varied, the induction method can be applied by changing the coil position in such a way that the total flux linked with the coil changes by a defined amount. A simple arrangement consists of a coil with flexible leads, Fig. (3-1)2, which is moved from a region where the field strength is practically zero into the magnetic field ($\Delta\Phi = \Phi$) or in the opposite direction ($\Delta\Phi = -\Phi$). The output voltage-time integral is observed. This method is particularly applicable for the measurement of fields in narrow gaps, using flat coils. Instead of

[1] R. Bernard and F. Davoine, *Compt. rend. acad. sci.*, **231**, 687 (1950).
[2] K. S. Lion, *Funk*, **5**, 431 (1928).

moving the coil in or out of the field, it can be rotated through an angle of 90° ($\Delta\Phi = \Phi$) or 180° ($\Delta\Phi = 2\Phi$).

Both methods are applicable for fields as small as $10^{-2}$ oersted. For smaller fields the coil dimensions become unwieldy or the output voltage becomes too small for convenient measurement.

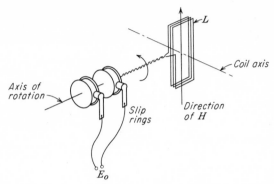

FIG. (3-1)3. Rotating search coil with slip rings for a-c output.

The continuous conversion of magnetic fields into electric signals is accomplished with an arrangement such as that shown schematically in Fig. (3-1)3. If the coil $L$ is placed in a uniform magnetic field $H$ and is rotated continuously with the angular frequency $\omega$, a sinusoidal voltage will appear at the output with an rms value

$$E_o = \frac{1}{\sqrt{2}} n a \mu H \omega$$

The ends of the coil are connected to slip rings, and contact is made with brushes. The axis of rotation must be perpendicular to the field direction. The rms value or peak value of the output is independent of the direction of the magnetic field normal to the axis of rotation. However, the phase angle between the instantaneous value of the a-c voltage output and the position of the rotating coil furnishes information about the field direction; the instantaneous voltage is a maximum when the coil plane is

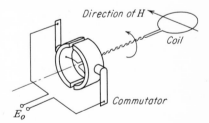

FIG. (3-1)4. Rotating search coil with commutator for d-c output.

parallel to the field lines. If a d-c output is desired, a rectifying commutator can be used as shown in Fig. (3-1)4. This arrangement permits a relatively simple determination of the field direction if the position of the contact brushes can be varied; the output is a minimum when the position of the brushes is in the direction of the magnetic field. An instrument built on this basis is described by Peters.[1] Its most sensitive range is 0.04 to 2 oersteds, its accuracy 1 to 2 per cent.

The methods illustrated in Figs. (3-1)3 and (3-1)4 are applicable for field strengths above about $10^{-2}$ oersted. If the fields are not

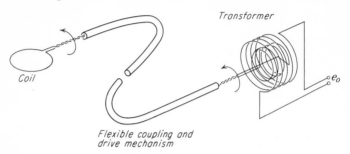

Fig. (3-1)5. Rotating search coil with inductive coupling to the output circuit.

homogeneous, the output will not be purely sinusoidal and will contain higher harmonics, so that an absolute determination of the field strength is difficult. Errors are likely to arise from brush noise, contact potentials, and thermoelectrical potentials at the brushes. The use of silver or siiver-graphite brushes lubricated with a thin layer of petroleum jelly has been recommended.

A method to overcome contact errors has been developed by Wills[2] and is illustrated in Fig. (3-1)5. The rotating coil is driven by a flexible cable; the connections from the coil are led to the primary winding of a transformer which rotates within a stationary secondary coil. The output from the secondary coil is measured. The loss of signal level due to leakage of the magnetic field of the transformer is considerable.

Errors may arise from variation of speed of the driving motor, from mechanical instability, such as oscillation of the coil and the suspension during rotation, and, in particular at low field strengths, from stray fields from the driving motor. Distances up to two meters between the motor and the pickup coil have been employed to reduce this source of error. Variations with time of the measured field

[1] W. A. E. Peters, *Elektrotech. Z.*, **71**, 193 (1950).
[2] M. S. Wills, *J. Sci. Instr.*, **29**, 374 (1952).

strength can be observed, but such times must be large compared with the period of rotation. The sensitivity varies widely with the construction and is in general between 0.1 and 10 mV/oersted. An accuracy of better than 0.1 per cent can be obtained.

The general requirements for the conversion of magnetic fields into electric voltages by means of oscillating coils are expressed in Eqs. (2) and (5a, b, and c). In some oscillation transducers, the angle between the field lines and the coil axis changes with time or is never

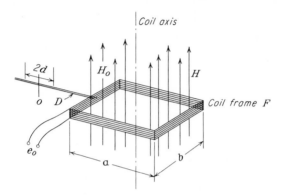

FIG. (3-1)6. Oscillating search coil, schematic diagram [*from* J. Grosskowski, *J. Sci. Instr.*, **14**, 335 (1937)].

zero degrees, so that the cos α term of Eq. (2) must be included in Eq. (5a, b, or c). The oscillation of the coil must produce a change of flux linked by the coil. The movement of a coil along its axis or perpendicular to its axis in a homogeneous field does not produce such a change of flux.

Figure (3-1)6 shows an example of an oscillating-coil transducer. A frame $F$ of a width $b$ is exposed at one end to a magnetic field with perpendicular component $H$. The field in the remaining part of the coil is $H_0$. ($H_0$ is small compared to $H$, or it is zero.) The coil vibrates, driven by a mechanical linkage $D$, in the direction of the arrow. The rms voltage appearing at the terminals is

$$E_o = \frac{2\pi}{\sqrt{2}} fdbn\mu(H - H_0)$$

where $d$ is the oscillation amplitude (measured from the middle position to one side), and $f$ is the oscillation frequency. The oscillation amplitude should be small compared to the homogeneous zone of the magnetic field.

A different kind of oscillating coil is shown in Fig. (3-1)7. The frame is placed in a magnetic field in the direction indicated and oscillates through an angle $2\alpha$. The output from the coil is

$$E_o = \frac{2\pi}{\sqrt{2}}\, anf\alpha\mu H$$

where $a$ is the area of the coil, $n$ the number of turns, $f$ the frequency of oscillation in cycles per second, $\alpha$ the amplitude of rotation in radians, and $H$ the component of the field strength perpendicular to both the axis of oscillation and the zero position of the coil axis. The

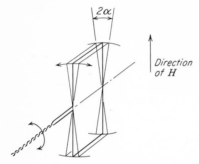

FIG. (3-1)7. Search coil for rotary oscillation.

output is not purely sinusoidal but contains the second harmonic, even for small angles of oscillation (1 to 4°). Klemperer and Miller[1] have measured field strengths of the order of magnitude of the earth's field with an accuracy of 1 per cent; the sensitivity of their instrument is $1.2 \times 10^{-3}$ volt/oersted.

Because the coil measures only one component of the field, as described above, two coils mounted perpendicularly upon each other can be used to measure two components of the field simultaneously without changing the position of the apparatus.[2]

Figure (3-1)8 shows a long coil of many turns. One end is in a region $A$ where the field strength parallel to the coil axis is $H_A$; the other end $Z$ is well into a region of zero field strength, and the field strength may fall off in any arbitrary fashion between $A$ and $Z$. The coil is oscillated in the direction of the axis with the amplitude $d$, and the frequency $f$. The induced voltage arises from the changing number of flux linkages and is described by Eq. (5c). It has the instantaneous value

$$e_1 = \frac{na}{l}\, \mu H_A d 2\pi f \cos(2\pi f t)$$

where $l$ is the length of the coil and $n$ the total number of turns.

[1] O. Klemperer and H. Miller, *J. Sci. Instr.*, **16**, 121 (1939).

[2] J. Kuntziger and P. Luonn, *Bull. Sci. AIM* (*Bel.*), **18**, 319 (1938); H. Neumann, *Arch. tech. Messen*, V 391-5, September, 1939.

If a field gradient $dH_A/dx$ exists at the end $A$, an additional voltage $e_2$ will be induced.

$$e_2 = na\mu \frac{dH_A}{dx} \frac{d^2}{2l} 2\pi f \sin (4\pi f t)$$

In practice, $e_2$ is much smaller than $e_1$, and since it is of twice the frequency, it can be separated.

If a very short, flat coil is oscillated longitudinally at $A$, it will produce $e_2$ only, and there will be no signal due to $H_A$.[1]

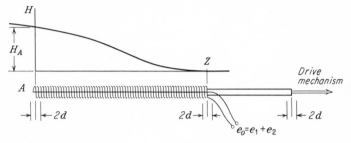

FIG. (3-1)8. Long search coil; one end $A$ extending into a magnetic field $H_A$; the magnetic field at the other end $Z$ is practically zero.

The method is suitable for measurements of flux densities in the order of 50 to 100,000 gauss and for field gradients from 1,000 to 100,000 gauss/cm. For a coil of 1,150 turns, 0.1 cm radius, 3.5 cm length, driven at a frequency of 1,000 cps with an amplitude of 0.001 cm, the sensitivities are

$$\frac{E_{1,\mathrm{rms}}}{\mu H_A} = 0.46 \times 10^{-6} \text{ volt/gauss}$$

$$\frac{E_{2,\mathrm{rms}}}{\mu \, dH_A/dx} = 2.3 \times 10^{-10} \text{ volt/(gauss/cm)}$$

The long coil transducer is suited for applications in measuring localized fields such as in magnetic lenses.[2]

Chapin[3] describes two oscillating magnetic-transducer systems which are driven by piezoelectric crystals and which measure an extraordinarily small area. The more sensitive system uses as a pickup coil a single wire loop of 1 mm length and a total travel of 0.154 mm. It is driven by a bimorphic crystal at a frequency of 872 cps. Its sensitivity is $3 \times 10^{-8}$ volt/gauss. The system is useful for

[1] J. F. Frazer et al., *Rev. Sci. Instr.*, **26**, 475 (1955).

[2] C. Fert and P. Gautier, *Compt. rend. acad. sci.*, **233**, 148 (1951).

[3] D. M. Chapin, *Rev. Sci. Instr.*, **20**, 945 (1949).

the measurement of flux density down to less than 1 gauss. The noise level is equivalent to 0.2 gauss.

*c. Special Search Coils.* COILS FOR USE IN INHOMOGENEOUS FIELDS. The determination of the local field strength in inhomogeneous fields requires a small search coil so that the field is practically homogeneous over the cross-sectional area of the coil. Such coils are inherently insensitive; the signal level at the output terminals of such coils is in general too small for measurement in magnetic fields of low field strength. The difficulty can be overcome by using the flux ball described by Brown and Sweer.[1]

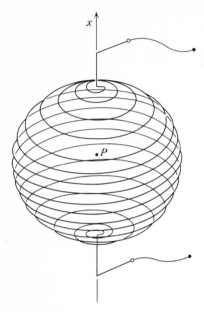

FIG. (3-1)9. Flux ball.

The method is based upon the fact that "the value of the magnetic field strength at a point is equal to the volume average of the field over the interior of any sphere centered at that point and not containing any source of the field" (Brown and Sweer). A sphere of this type surrounding the point $P$ is shown in Fig. (3-1)9. The net flux linked with the coil

$$\Phi = \tfrac{4}{3}\pi r^3 n B_{z,\text{center}}$$

is where $r$ is the radius of the sphere, $n$ the number of turns per unit of length measured along the axis, and $B_{z,\text{center}}$ is the $z$ component of the flux density at the center of the sphere if the axis is in the $z$ direction.

For increased sensitivity, the entire sphere is filled with windings (solid flux ball) made up of cylindrical coils of different lengths and diameters as shown in Fig. (3-1)10. Such a coil may be thought of as consisting of concentric spheres; the net flux linked with the solid flux ball is the sum of those linked with each sphere. The entire flux linked with the solid flux ball

$$\Phi' = \frac{4}{15}\,\pi\,\frac{1}{D_\rho D_z}\,r^4 B_{z,\text{center}}$$

[1] F. W. Brown, Jr., and J. H. Sweer, *Rev. Sci. Instr.*, **16**, 276 (1945).

where $D_\rho$ is the layer spacing, and $D_z$ is the spacing between adjacent turns (effective wire diameter).

A disadvantage of the flux ball is its physical dimensions, which are considerably larger than those of a flat coil, and which make it impossible to measure fields in narrow gaps or in close proximity of a magnetic pole.

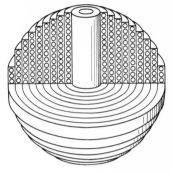

Herzog and Tischler[1] have shown that an ordinary cylindrical coil can be employed instead of the flux ball if the ratio of the length $L$ to the outer diameter $D$ has the value $L/D = 0.72$. The voltage-time integral at the output of this coil depends only upon the change of field intensity in the center of the coil. The authors found that in using coils of this shape the results of measurements in an inhomogeneous field are practically independent

Fig. (3-1)10. "Solid" flux ball, partly seen in cross section to show the arrangement of the single layers.

(within less than 1 per cent) of the size of the coil.

*d. Magnetomotive-force Transducer Coil.* A magnetic potential difference or magnetomotive force $M$ between two points $A$ and $B$ is defined as

$$M = \int_A^B H\,ds$$

it is independent of the direction and length of the path as long as $s$ does not surround a current-carrying conductor. The transformation of a magnetomotive force into an electric signal can be obtained, in analogy to the transformation of magnetic field strengths into electric signals, through the use of a special coil, the "magnetomotive-force meter."[2]

The instrument consists of a long coil (about 1 m in length), as shown in Fig. (3-1)11. Two layers of fine wire are wound on a band or belt of a flexible plastic material, both in the same direction, as shown schematically in Fig. (3-1)12. For practical reasons, connections are made to the center of the coil.

If the two ends of the coil are brought to two points in a magnetic field between which a magnetomotive force $M$ exists, and if the field

[1] R. F. K. Herzog and O. Tischler, *Rev. Sci. Instr.*, **24**, 1000 (1953).

[2] W. Rogowsky and W. Steinhaus, *Arch. Elektrotech.*, **1**, 141 (1912); for a summarizing review and references, see H. Neumann, *Arch. tech. Messen*, J 64–1, May, 1934.

is turned on or off or if the coil is removed from the two points, a voltage-time integral will arise between the output terminals, which is

$$\int E \, dt = \frac{\mu n a}{l} M$$

The number of turns $n$, the cross-sectional area $a$ of each turn, and

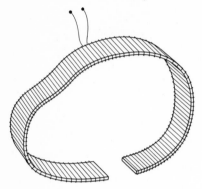

FIG. (3-1)11. Search coil for measurement of magnetomotive force.

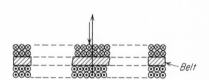

FIG. (3-1)12. Cross section of the search coil illustrated in Fig. (3-1)11.

the length $l$ can be measured separately, or the coil can be calibrated empirically.

## 3-12. SYSTEMS BASED UPON MAGNETIC EFFECTS ON MOVING CHARGES

*a. Rotating Disk.* A magnetic transducer based upon the effect of the magnetic field upon the electrons in a moving-solid conductor is

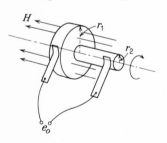

FIG. (3-1)13. Rotating disk ("unipolar" generator) as a transducer for magnetic field strength.

illustrated in Fig. (3-1)13. It consists of a rotating metal disk (copper, aluminum) with contact brushes applied to the axis and the periphery. The conduction electrons in the metal disk are moving with the disk in circular paths. Under the influence of the magnetic field, the electrons are deflected in a direction perpendicular to the mechanical movement and the magnetic field, thus causing a potential difference between the center of the disk and the periphery. With the notations of Fig. (3-1)13, the (d-c) output voltage is

$$e_o = \pi(r_1^2 - r_2^2)f\mu H$$

where $f$ is the frequency of rotation in revolutions per second. The sensitivity of the method is low.[1]

*b. Induction in Moving Conductive Liquids.* Instead of a moving solid conductor, a conductive liquid flowing through a magnetic field can be used for the construction of a magnetoelectric transducer.[2]

Figure (3-1)14 shows a schematic diagram of the mercury jet magnetic transducer of Kolin. A stream of mercury passes through an insulating tube with the average velocity $\bar{v}$. Two fixed, diametrically opposite electrodes $A$ and $B$ are inserted in the tube, their small end surfaces flush with the inner walls of the tube. Instead of mercury, any other conductive liquid can be used (e.g., salt water).[3] If the transducer is exposed to a magnetic field $H$ in the direction perpendicular to the plane formed by the tube axis and the line connecting the electrodes, a voltage appears between the electrodes $A$ and $B$ of a magnitude

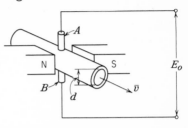

FIG. (3-1)14. Mercury-jet magnetic transducer [*from A. Kolin, Rev. Sci. Instr.*, **16**, 209 (1945)].

$$E_o = d\bar{v}\mu H$$

where $d$ is the distance between the electrodes, i.e., the internal diameter of the tube. A mathematical analysis of the transducer principle is presented by Thürlemann,[4] with consideration of electrical eddy currents originating in the conductive liquid. The output voltage is proportional to the average flow velocity for both laminar and turbulent flow and is independent of the velocity distribution over the cross section of the tube.

Since the mercury is flowing continuously, fast field changes can be followed and the method is applicable to the measurement of time varying or a-c fields of a frequency up to several kilocycles. Field directions and field gradients can be measured by the use of further pairs of electrodes.

The absolute accuracy depends on the accuracy with which $d$, the separation of the electrodes, and $\bar{v}$ can be determined. The velocity can be measured with an accuracy of 0.1 per cent. With reasonable

[1] Cited by H. Neumann, *Arch. tech. Messen*, V 391–5, September, 1939, and V 391–4, August, 1940; also L. W. McKeehan, *J. Opt. Soc. Am.*, and *Rev. Sci. Instr.*, **19**, 213 (1929).

[2] First proposed by A. Leduc, *J. Physique*, **46**, 184 (1887).

[3] See B. Thürlemann, *Helv. Physica Acta*, **14**, 383 (1941).

[4] Thürlemann, *ibid.*

physical dimensions, the sensitivity is of the order of 0.4 microvolt/ oersted. The instrument is applicable to the measurement of field strengths above 1 oersted.

*c. Magnetron Tubes.* A magnetron, shown in Fig. (3-1)15, consists of an electron-emitting cathode $C$ surrounded concentrically by a cylindrical anode $A$. Electrons move from the cathode to the anode under the action of an applied voltage $E_b$ and form a current $I$ in the external circuit. The path of the electrons is radial, as shown in

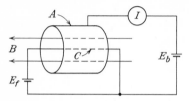

FIG. (3-1)15. Magnetron tube, schematic diagram and circuit.

FIG. (3-1)16. Electron paths in a magnetron.

Fig. (3-1)16, path 1. If a magnetic field $H$ is applied in the direction of the tube axis, the electrons will travel in a curved path (path 2), and if the field strength is increased, a condition will be reached in which no electrons reach the anode (path 3) and the current diminishes to zero. The condition at which this occurs depends upon the intensity of the magnetic field $H$ and the applied plate voltage $E_b$. The critical magnetic field strength at which current cutoff occurs is

$$H_c = \frac{1}{\mu r_a[1 - (r_c^2/r_a^2)]} \sqrt{8 \frac{m}{e} E_b} \tag{1}$$

where $e$ is the charge and $m$ the mass of the electron, $r_c$ the radius of the cathode, and $r_a$ the radius of the anode. If the diameter of the cathode is small compared to that of the anode, the term $1 - (r_c^2/r_a^2)$ in Eq. (1) can be neglected.

In practice, the critical magnetic field strength is not accurately defined. The current in Fig. (3-1)17 decreases with a finite slope owing to tube asymmetries, variations of the initial velocity of the electrons, space charge (occurring in particular at low plate voltages), fringe effects at the ends of the cylinder, and, in directly heated tubes, differences of the effective plate voltage along the axis. Empirical calibration of the transducer is required for accurate results.

Three basic methods for the operation of the magnetron transducer are available:

1. The plate voltage $E_b$ is kept constant, as indicated by the lines

*aa* in Figs. (3-1)17 and (3-1)18, and the current $I_b$ is measured.[1] The method is applicable for field strengths below 20 oersteds and for small variations of field strength. The sensitivity is high, of the order of 20 mA/oersted. Feedback circuits for the increase of sensitivity and stability are described by Rössiger.[2]

2. The current $I_b$ is kept constant by variation of the plate voltage $E_b$, as shown by the line *bb* in Fig. (3-1)18. The voltage $E_b$ is

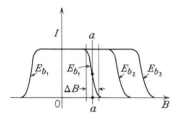

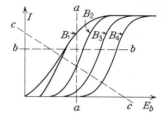

Fig. (3-1)17. Characteristics of a magnetron illustrating the plate current $I$ as a function of the magnetic field strength $B$ for different plate voltages $E_b$.

Fig. (3-1)18. Plate current-plate voltage characteristics of a magnetron; parameter, magnetic field strength.

measured. The method is primarily applicable for medium field strength and for large variations of field strength. A direct-reading instrument based upon this method is described by Gundlach.[3] The range extends from about 200 to 1,200 gauss.

3. A resistance is inserted in the plate circuit; the applied voltage $E_b$ is kept constant, and the current $I_b$ or the voltage across the resistor or the magnetron tube is measured, as shown by line *cc* in Fig. (3-1)18. This method, first proposed by Hull, is applicable for small or medium fields, depending upon the size of the resistor.

The magnetron transducer has the advantage that no moving parts and no variation of the measured field are required. The inertia of the system is negligible up to $10^{-8}$ sec. The probe can be made fairly small and has no effect upon the measured field. Weinzierl[4] has used anode cylinders with a diameter as small as 0.5 mm sealed in glass tubes of 6 mm diameter. Ferromagnetic materials, such as nickel, must be avoided in the construction of the tube. If operated under temperature-limited conditions, the filament temperature

[1] A. W. Hull, *Phys. Rev.*, **22**, 279 (1923); M. Rössiger, *Z. Physik*, **43**, 480 (1927), and *Z. Instrumentenk.*, **49**, 105 (1929).

[2] Rössiger, *ibid*.

[3] F. W. Gundlach, *Arch. tech. Messen*, V 391–3, July, 1938.

[4] P. Weinzierl, *Rev. Sci. Instr.*, **21**, 492 (1950).

must be kept very constant to keep the tube from drifting. At high field strengths and high plate voltages the operation tends to become unstable and oscillations of very high frequencies may set in.

*d. Electron-beam Magnetometer.* The observation of the magnetic deflection of a long, focused electron beam furnishes a sensitive method for the detection and measurement of very small magnetic fields and field variations in the order of $10^{-6}$ oersted.[1]

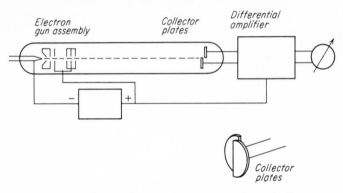

Fig. (3-1)19. Electron-beam magnetometer, schematic diagram.

The arrangement is shown schematically in Fig. (3-1)19. It consists of a vacuum tube with an electron-gun assembly (electron-emitting, accelerating, and focusing system) at one end. At the other end at a distance of 1 m are two semicircular collector plates. The collector plates are overlapping but electrically separated. An electron beam of small cross section strikes both plates, and if the beam is symmetrical, the electron currents to both plates are equal. The current is measured by a differential d-c amplifier; the difference furnishes a measure of the beam deflection caused by a magnetic field $H$, in Fig. (3-1)19 assumed to be perpendicular to the drawing plane, and hence can be used for the determination of the field.

The deflection of the electron beam is

$$y = 0.15 k B l^2 \frac{1}{\sqrt{E}}$$

where $y$ is measured in centimeters; $B$ in gauss; $l$, the length of the beam, in centimeters; and $E$, the accelerating voltage, in volts. The constant $k$ is determined by the optics of the system (order of magnitude 0.5 to 0.7).

[1] L. Marton, L. B. Lederer, and J. W. Coleman, *Natl. Bur. Standards Rept.* 2381, March, 1952.

The sensitivity of this method is very high, permitting a measurement of magnetic fields of the order of $8 \times 10^{-6}$ gauss with a signal-to-noise ratio of 9.7:1.

The experimental difficulties are considerable. The accelerating voltage, which is of the order of 600 to 1,500 volts, must be kept constant to 1 part in $10^4$. The output current is of the order of $10^{-8}$ amp; for the desired sensitivity the differential amplifier must be built to detect differences in current of about $10^{-12}$ amp.

*e. Hall-effect Transducers.* A plate of an appropriate conductor or semiconductor such as bismuth or germanium is connected to a voltage source as shown in Fig. (3-1)20, so that a constant current $I$ passes through it. Two contacts

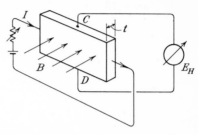

Fig.   (3-1)20.   Hall-effect   magnetic transducer.

$C$ and $D$ are applied to equipotential points, usually in the middle of the sides. If a magnetic field $B$ traverses the plate in a direction perpendicular to the current vector $I$ and the connecting line $CD$, it will tend to deflect the moving charges toward the top or the bottom of the plate, depending upon the sign of the charges. Thus, a potential difference, the Hall voltage, will develop between the electrodes $C$ and $D$, which is

$$E_H = K_H \frac{IB}{t}$$

where $E_H$ is expressed in volts, $I$ in amperes, $B$ in gauss, and $t$, the thickness of the plate, in centimeters. $K_H$ is the Hall coefficient in volt-centimeters per ampere-gauss.

The polarity of the output voltage depends upon the direction of the current and of the magnetic field, as well as upon the sign of the Hall coefficient. The dimensions of the plate are not critical. Practical values are 1.2 cm length, 0.4 cm width, and 0.05 cm thickness. The Hall voltage varies somewhat with the ratio of width $W$ to length $L$ of the plate.[1] A ratio $W/L = \frac{2}{3}$ is recommended.

The Hall coefficient $K_H$ depends upon the material of the plate, as shown in Table 13. The value of $K_H$ for bismuth, silicon, and tellurium is higher by a factor of about $10^5$ than that of most other metals; the value for germanium by a factor of about $10^8$. The values of $K_H$ for different substances vary with the amount of impurities. Plates

---

[1] A. von Ettinghausen and W. Nernst, *Wien. Ber.*, **94**, 644 (1886).

must be calibrated individually. The Hall coefficient changes with the temperature and with the magnetic field strength. At room temperature and at a field strength of 1,000 gauss, a temperature increase of 10°C causes an increase of the Hall coefficient for germanium by about 4 per cent, and for bismuth by about 10 per cent.

TABLE 13. HALL COEFFICIENTS FOR DIFFERENT MATERIALS

| Material | Field strength, gauss | Temp., °C | $K_H$ |
|---------|----------------------|-----------|-------|
| As | 4,000–8,000 | 20 | $4.52 \times 10^{-11}$ |
| C | 4,000–11,000 | Room | $-1.73 \times 10^{-10}$ |
| Bi | 1,130 | 20 | $-1 \times 10^{-8}$ |
| Cu | 8,000–22,000 | 20 | $-5.2 \times 10^{-13}$ |
| Fe | 17,000 | 22 | $1.1 \times 10^{-11}$ |
| $n$-Ge | 100–8,000 | 25 | $-8.0 \times 10^{-5}$ |
| Si | 20,000 | 23 | $4.1 \times 10^{-8}$ |
| Sn | 4,000 | Room | $-2.0 \times 10^{-14}$ |
| Te | 3,000–9,000 | 20 | $5.3 \times 10^{-7}$ |

The influence of the magnetic field strength upon the Hall coefficient is different for various substances. The Hall voltage in bismuth rises first with increased field strength, reaches a maximum at 5 to 8 kilogauss (depending upon temperature), and falls off to zero or negative values between 10 and 20 kilogauss.[1] Anomalies in the behavior of bismuth at low magnetic field strengths limit the applications at fields below 0.3 gauss.[2]

The output voltage is in the milli- and microvolt region. The output impedance depends upon the resistivity of the material and the dimensions of the probe plate; the resistivity of those substances which have a high Hall coefficient varies considerably with the temperature and the magnetic field strength. A transducer system based upon the Hall effect is described by Peukert.[3] The author uses a plate of bismuth, $6.8 \times 2.8 \times 0.1$ cm and a current $I$ of 1.5 amp. Pearson[4] describes a transducer using n-type germanium with a range up to 20,000 gauss. The deviation from linearity (as determined by comparison with a rotating coil meter) is $\pm 2$ per cent up to 8,000 gauss and $-9$ per cent at 20,000 gauss. The output impedance varies between 140 and 170 ohms.

*f. Magnetoresistive Transducers.* The resistivity of metals at low

[1] G. Bublitz, *Arch. tech. Messen*, V 391–2, May, 1938.

[2] P. H. Craig, *Phys. Rev.*, **27**, 772 (1926).

[3] W. Peukert, *Elektrotech. Z.*, **31**, 636 (1910).

[4] G. L. Pearson, *Rev. Sci. Instr.*, **19**, 263 (1948).

temperature changes if the metal is exposed to a magnetic field.  The effect is noticeable in some metals at room temperature and is strongest in bismuth; the resistivity of bismuth in strong magnetic fields (of the order of 20,000 oersteds) can rise to twice that measured at zero field strength.  This phenomenon forms the basis of the magneto-resistive transducer system shown in Fig. (3-1)21.

A thin wire of bismuth is wound in the form of a flat bifilar spiral of about $\frac{1}{2}$ to 3 cm diameter.  The spiral is cemented between two thin

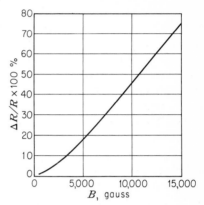

FIG. (3-1)21. Magnetoresistive transducer (bismuth spiral).

FIG. (3-1)22. Calibration curve of a magnetoresistive transducer.

sheets of mica.  In general the thickness of this probe does not exceed 1 mm.  The ends of the wire are soldered to two copper bands, insulated from each other, which serve both as connecting elements and as a handle.  The bifilar arrangement eliminates induction effects when the coil is moved in or out of a magnetic field or when the field changes.

A typical transfer characteristic is shown in Fig. (3-1)22.[1]  In the linear part of the characteristic the resistance increases by about 5 per cent for an increase of the magnetic field of 1,000 oersteds.  Below 400 oersteds the characteristic shows an irregular and anomalous behavior.[2]

The resistance variation of bismuth in magnetic fields shows a transient aftereffect which is not fully understood.  The effect depends upon the magnetic field strength and the current through the bismuth probe.[3]  Because of this effect the bismuth spiral should

[1] G. Bublitz, *Arch. tech. Messen*, V 391–2, May, 1938.

[2] G. K. T. Conn and B. Donovan, *J. Sci. Instr.*, **28**, 7 (1951).

[3] See, for instance, P. P. König, *Ann. Physik*, **25**, 921 (1908), and T. Heurlinger, *Physikal. Z.*, **17**, 221 (1916).

be used only for the measurement of constant (d-c) magnetic fields, and the resistance should be measured with direct current.

The response characteristics of bismuth transducers depend strongly upon the chemical composition (i.e., impurities) and the method of preparing and handling the wire. Using very pure bismuth, compressing the wire while at a temperature close to the melting point and annealing the spiral are recommended.[1] The response also depends somewhat upon the geometric form of the spiral and the direction of the magnetic field. The bismuth wire shows, in general, an anisotropic behavior. Conn and Donovan[2] describe a method of preparing a bismuth wire by drawing bismuth in glass capillaries, dissolving the glass, and forming the wire into a loop. This method reduces the anisotropy. The magnetic sensitivity—slope, $(\Delta R/R)/\Delta B$—of thin bismuth films is very small but can be slightly increased by repeated heating and cooling.[3]

The characteristic of the prepared probe changes considerably with temperature. In the linear part of the characteristic and at room temperature, a change in temperature of 1°C causes a change in resistance of about 0.4 per cent (corresponding to a change of the magnetic field of 80 oersteds). The sensitivity increases considerably at lower temperature. For this reason and since the thermoelectric effect of bismuth against other metals is very high and likely to introduce errors, operation of the transducer at constant temperature and measurement of the resistance at low current (to avoid heating) are required. With adequate provisions for constant temperature or with temperature-effect compensation, the transducer can be used for magnetic fields from 500 oersteds up to the highest field strengths with an accuracy of about 1 per cent. Smith[4] describes a bridge in which two bismuth spirals of approximately 30 ohms are arranged in opposite bridge arms. For temperature compensation the two other bridge arms are made from nickel which has approximately the same resistance-temperature coefficient as bismuth. The sensitivity $S = \Delta i_o/\Delta B$, i.e., the change of current in the bridge diagonal indicated by the meter divided by the change of the magnetic flux density in the bismuth plate, is of the order of 1 mA per 1,000 gauss.

For a theoretical summarizing paper and references concerning other materials showing magnetoresistive effect, see J. P. Jan, "Galvanomagnetic

[1] G. Bublitz, *loc. cit.*

[2] G. K. T. Conn and B. Donovan, *loc. cit.*

[3] J. A. Becker and L. F. Curtis, *Phys. Rev.*, **15**, 457 (1920), and F. K. Richtmyer and L. F. Curtis, *Phys. Rev.*, **15**, 465 (1920).

[4] G. S. Smith, *Trans. AIEE*, **56**, 441 (1937).

and Thermomagnetic Effects in Metals," in "Solid State Physics," pp. 1ff and sec. 10, p. 30, F. Seitz and D. Turnbull (eds.), Academic Press, Inc., New York, 1957.

### 3-13. TRANSDUCERS BASED ON PERMEABILITY VARIATION

The magnetic permeability $\mu(= dB/dH)$ of ferromagnetic materials varies with the magnetization and is, therefore, a function of the external magnetic field to which a ferromagnetic body is exposed. Various transducer systems are based upon this phenomenon; they differ by the means by which the change in permeability is detected.

*a. Inductance Variation.* Boucke[1] describes a system consisting of a coil with a ferromagnetic core; the inductance of the coil is a function of the permeability of the core, and it changes with the external magnetic field. A ferromagnetic core of high permeability would cause a field deformation and introduce a considerable error in the measurements. This error is reduced to about 10 to 20 per cent by the use of a core of compressed powdered magnetic material (50 to 70 per cent iron powder: grain size, several microns, with a resinous binder). The most suitable core material shows an initial permeability between 10 and 12, which diminishes to about 2 to 3 in strong magnetic fields. A flat washer is formed of this material, and a toroidal

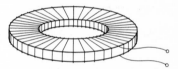

FIG. (3-1)23. Inductance-variation magnetic transducer.

coil is wound around it, as shown in Fig. (3-1)23. The final coil is about 1 mm thick and about 6 mm in diameter; it is electrostatically shielded and has an inductance of several microhenrys.

The transfer characteristic, i.e., the fractional change of inductance ($\Delta L/L$) as a function of the externally applied field, is nonlinear; in the range from zero to 12 oersteds the value of $\Delta L/L$ increases slightly; beyond this range it decreases in the form of an S curve. The fractional change of inductance is about 10 per cent for a field variation from zero to 24,000 oersteds. The sensitivity changes with the direction of the magnetic field with respect to the coil. The core material does not show any hysteresis.

Using the most sensitive method for the measurement of the inductance variation $\Delta L$ (rf beat-frequency method), a magnetic-field variation of 0.01 per cent can be observed in the range between 300 and 2,000 oersteds. The smallest absolute field strength to be measured is about 1 oersted. Empirical calibration is required.

[1] H. Boucke, *Arch. tech. Messen*, V 391–1, June, 1937.

Because of the large residual field deformation caused by the probe, the method lends itself primarily to relative measurements.

*b. Time Variation of Magnetic Permeability.* A variation of the above method is illustrated in Fig. (3-1)24. The transducer system contains a fine wire $M$ of Mumetal[1] (a ferromagnetic alloy with a permeability which changes strongly with the magnetizing field). The wire is connected in a loop and arranged within a coil $L$. An a-c current of audiofrequency passes through the wire $M$ and produces a circular field in and around the wire. This field changes the permeability of the Mumetal wire, but because of its direction it will not, by itself, induce a voltage in the coil $L$. If now an external field $H_0$ is applied in the direction of the coil axis, it will produce a magnetic induction $B_z = \mu H_0$. Since $\mu$ is a time varying magnitude, $B_z$ varies with time and causes a voltage to be induced in the coil $L$. The frequency of this voltage contains primarily the even harmonics of the applied audiofrequency. Experimental calibration is required. If the instrument is to be used as a zero device, a d-c current can be sent through the coil $L$ which compensates the external field $H_0$. Palmer[2] describes an instrument based upon this principle, which is capable of measuring field strengths from $2 \times 10^{-3}$ ($10^{-5}$ with great care) to 50 oersteds. The accuracy is better than 0.1 per cent. The probe is about 4 cm long and 0.5 cm in diameter. A similar device with two coils is described by Gregg.[3] The permalloy wire is 0.3 cm long and has a diameter of 0.25 mm. The finished probe is 0.3 cm long and has a diameter of 0.2 cm.

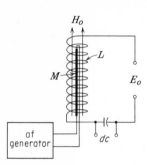

FIG. (3-1)24. Magnetic transducer based upon the variation with time of a Mumetal wire $M$; $L$, pickup coil.

*c. "Even-harmonic" Method.* Higher sensitivity and higher accuracy than can be achieved with the methods described above can be obtained with the system shown schematically in Figs. (3-1)25

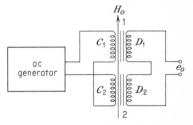

FIG. (3-1)25. Magnetic transducer based upon the "even-harmonic" method: $C_1$ and $C_2$ primary, $D_1$ and $D_2$ secondary windings of a transformer.

---

[1] Mumetal, trade name, Allegheny Ludlum Steel Corporation.

[2] T. M. Palmer, *Proc. Inst. Elec. Engrs.* (*London*), **100**, (II) 545 (1953).

[3] E. C. Gregg, *Rev. Sci. Instr.*, **18**, 77 (1947).

and (3-1)26. A sinusoidal voltage from an a-c generator is applied to two probe coils $C_1$ and $C_2$. The coils contain ferromagnetic cores of a material which shows high permeability and saturation at low magnetizing field strength (permalloy). The current is high enough so

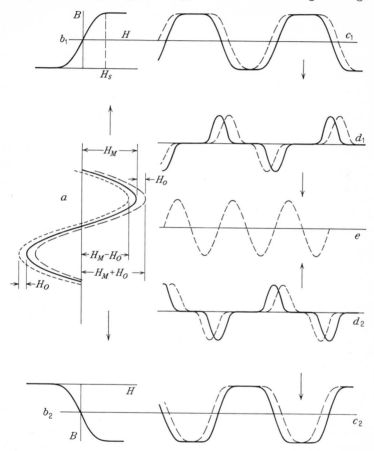

FIG. (3-1)26. Voltage–time diagram of the transducer shown in Fig. (3-1)25: solid lines, without external magnetic field $H_0$; dashed lines, with field $H_0$; (a) magnetizing field; (b) magnetization curves; ($c_1$) field in core $c_1$; ($c_2$) in core $c_2$; ($d_1$) voltage induced in coil $d_1$; ($d_2$) in coil $d_2$; (e) output after filtering.

that saturation occurs. Voltages of equal magnitude but opposite direction are induced in the secondary coils $D_1$ and $D_2$. These voltages are not sinusoidal and contain large percentages of the odd harmonics. The coils are connected in series, so that the total output is zero. If now a (small) external field $H_0$ is applied in the direction of the core axis, the two cores will be magnetized in the same direction by the external field and in the opposite direction by the magnetizing a-c current. As shown in Fig. (3-1)26, $d_1$ and $d_2$, the voltages induced in the secondary coils $D_1$ and $D_2$, will be phase-shifted in different directions so that an output voltage containing primarily the second harmonic (and higher even harmonics) of the excitation frequency will appear at the output terminals, as shown in Fig. (3-1)26e. This voltage can be filtered, amplified, and rectified in subsequent stages. The output voltage of the second harmonic is

$$E_{o,\mathrm{rms}} = 44 \times 10^{-9} nafh\mu_r H_0$$

where $n$ is the number of turns of the coils $C_2$ and $D_2$, $a$ the cross-sectional area of the ferromagnetic core in square centimeters, $\mu_r$ the (effective) permeability of the core, $f$ the frequency of the a-c generator, and $H_0$ the externally applied field. The factor $h$ depends upon the ratio of magnetizing field strength $H_M$ to the field strength $H_S$ at which saturation occurs:

$$h = \frac{4}{\pi} \sin \frac{2H_S}{H_M}$$

The method can be used for very small field strengths from $5 \times 10^{-4}$ oersted[1] to above 10 oersteds. Adams, Dressel, and Towsley[2] describe an instrument for the range of $10^{-3}$ to $100$ oersteds. The transfer function (output voltage versus external field strength) is approximately linear; only at high field strength does the curve flatten out, and at very low field strength (at about $10^{-5}$ oersted) does an output voltage remain which does not diminish to zero at zero oersteds. Reversal of the field direction reverses the phase but not the magnitude of the output voltage. If one of the cores is inverted so that not only the fundamental frequency but also the even harmonics cancel, the arrangement can be used for the measurement of magnetic-field differences. The response time depends upon the frequency of the excitation voltage. For frequencies in the order of 1 to several kilocycles per second, magnetic fields varying within more than 0.01 sec can be measured. No errors due to residual flux are encountered.

The sensitivity is of the order of 1 volt/oersted; the output impedance usually several hundred ohms. A summarizing report on systems of this type, including references, has been published by Kühne.[3]

[1] V. Vaquier, R. F. Simons, and A. W. Hull, *Rev. Sci. Instr.*, **18**, 483 (1956).
[2] G. D. Adams, R. W. Dressel, and F. E. Towsley, *Rev. Sci. Instr.*, **21**, 69 (1950).
[3] R. Kühne, *Arch. tech. Messen*, V 392–1, August, 1952.

d. *"Peaking Strips."* The arrangement is shown schematically in Fig. (3-1)27. It consists of a wire probe $M$ of a ferromagnetic material, such as permalloy, which exhibits a hysteresis loop, as shown in Fig. (3-1)28. The wire probe is surrounded by three coils. In Fig. (3-1)27 these are shown separately; in practice the coils should be long and overlapping. The coil $C_1$ is connected to an a-c source and produces an alternating magnetic field varying periodically between $-H_s$ and $+H_s$. Every time the field passes in the positive direction

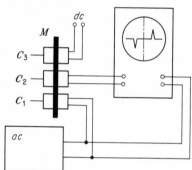

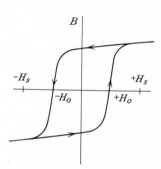

FIG. (3-1)27. "Peaking-strip" magnetic transducer.

FIG. (3-1)28. $HB$ characteristic of the peaking strip.

through $+H_0$ and in the negative direction through $-H_0$, the magnetization changes sharply and causes a positive or negative pulse in the second coil $C_2$. These pulses can be detected by means of a cathode-ray oscilloscope. If the a-c source which supplies the coil $C_1$ is also applied to the X-axis deflection system of the oscilloscope, the pattern will be symmetric. If now an external magnetic (d-c) field is applied, the two peaks appear asymmetric until the effect of the external field upon the wire probe is compensated for by means of a d-c current applied to the coil $C_3$. The magnetic field strength produced by the coil $C_3$ is equal and of opposite direction to the external field; the instrument does not require an empirical calibration.

The instrument is useful for the measurement of magnetic fields in a range from 0.1 to above 500 oersteds. The upper limit is set by the maximum field strength attainable in the coil $C_3$. If this coil $C_3$ is water-cooled, an upper limit of 5,000 oersteds may be reached. The smallest deviations which may be detected are of the order of 0.02 oersted. The limitation is due to thermal effects and to mechanical instability of the system. The output is, in general, of the order of a few millivolts; the output impedance 10 to several hundred ohms.

The method is applicable only to steady fields or to time-varying fields of a frequency which is low compared to the frequency of the a-c current supplied to the coil $C_1$. The more uniform the field strength, the sharper are the pulses induced in $C_2$; the method furnishes poor results in inhomogeneous fields.

The dimensions of the ferromagnetic probe $M$ are not critical; the ratio of length to diameter should be large to obtain narrow and well-defined pulses. The magnetic properties of most suitable ferromagnetic materials are quite dependent upon temperature and mechanical deformation; they vary with the magnitude of the applied field and the frequency of the a-c generator. Kelly[1] recommends the use of a molybdenum-permalloy wire of 5 cm length and 0.005 cm diameter, annealed in a hydrogen atmosphere inside a quartz tube and sealed after annealing.

### 3-14. NUCLEAR MAGNETIC RESONANCE SYSTEMS

*a. Nuclear-absorption Method.* Protons and neutrons possess an intrinsic magnetic moment, and hence atomic nuclei have, in general, a magnetic moment of a magnitude which depends upon the number and the arrangement of the protons and neutrons in the nucleus. If a nucleus with a magnetic moment is placed in a magnetic field, the field will exert a force on the nucleus. Because of the quantization of the nuclear angular momentum which always accompanies the magnetic moment, the nucleus can assume only certain orientations in a magnetic field, each orientation corresponding to a certain level of energy. In the simplest case of a proton ($H^1$) which has only two permitted orientations in the field, there will be two states of slightly different energy corresponding to the two orientations. The energy difference $\Delta E$ between the two energy levels is proportional to the magnetic field strength. Transition from the lower to the higher energy level is accomplished by an absorption of energy. The energy can be supplied by an electromagnetic field of frequency ("resonance frequency")

$$f_0 = \frac{\Delta E}{h}$$

where $h$ is Planck's constant. The energy $\Delta E$ is proportional to the external magnetic flux density $B$, and hence

$$f_0 = \gamma B \tag{1}$$

Values of $\gamma$ are: proton in $H_2O$, 4257.76 $\pm$ 0.1 cps/gauss; lithium[7],

[1] J. M. Kelly, *Rev. Sci. Instr.*, **22**, 256 (1951).

1654.61 ± 0.1 cps/gauss. Measurement of magnetic field strength can be accomplished, therefore, by the determination of the frequency at which a sample containing hydrogen or lithium absorbs energy. Using these substances and flux densities of $10^2$ to $10^4$ gauss, the nuclear-resonance frequency is in the range between $10^5$ and $10^8$ cps.

A schematic diagram of a transducer system based upon this nuclear-resonance phenomenon is shown in Fig. (3-1)29.

A sample $N$ of a salt or a solution of a substance containing $H^1$, for instance, or $Li^7$, is mounted within a coil $C$ and is brought into the magnetic field between the poles $P_1$ and $P_2$ of a magnet, so that the axis of $C$ is perpendicular to the magnetic field. The output from an rf oscillator is applied to the coil. The frequency of the oscillator is varied continuously through resonance, and the absorption of energy is observed with the measuring system $M$. The frequency at which a maximum of absorption occurs is measured, and the unknown field strength is computed from Eq. (1).

FIG. (3-1)29. Nuclear magnetic resonance system, basic diagram: $C$, coil; $N$, sample; $P_1$ and $P_2$ poles of magnet.

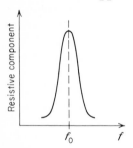

FIG. (3-1)30. Nuclear magnetic absorption curve.

Two different systems $M$ for the detection of the nuclear magnetic resonance frequency can be used. The measuring system $M$ can be constructed so that its output indicates the change in the resistive component of the complex coil impedance. In this case a curve with a maximum is found as shown in Fig. (3-1)30 (nuclear-resonance absorption). Or the measuring system $M$ furnishes an output which is proportional to the reactive component of the resonant circuit, so that a curve is found as shown in Fig. (3-1)31 (nuclear-resonance dispersion).

The method is practical for field strengths from about 300 to 20,000 gauss (for proton probes mostly from 300 to 8,000 gauss, for lithium probes from 2,000 to 20,000 gauss) corresponding to a frequency range from 1 to about 30 Mc. Field strengths have been measured down to the order of the earth's magnetic field.[1] The accuracy of the method is very high and is primarily determined by the accuracy

[1] M. E. Packard and R. Varian, *Phys. Rev.*, **93**, 941 (1954).

with which the frequency of the rf generator can be measured. Accuracies as high as 1 part in $10^4$ to $10^5$ have been obtained.

The nuclear magnetic resonance effect is very small and the resonance is very sharp. (The Q factor of a nuclear-resonance system is in the order of $10^6$). The requirements for the equipment to detect the resonance absorption and to measure the frequency are considerable.

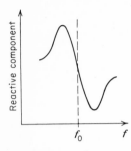

FIG. (3-1)31. Nuclear magnetic resonance dispersion curve.

A method to facilitate the detection of resonance is shown in Fig. (3-1)32. The basic arrangement is identical with that of Fig. (3-1)29, but a second field is applied (coils $D$) which is energized by a 25- to 60-cycle source. These coils produce an a-c magnetic field of approximately 5 to 30 gauss which is superimposed on the steady field and which periodically changes the resonant frequency of the probe. The a-c source is also connected to the horizontal deflection system of a cathode-ray oscilloscope; a signal which is proportional to the absorbed power is formed in the network $M$ and is applied to the vertical deflection system. Because the a-c field periodically changes the resonance frequency of the probe, a resonance curve will

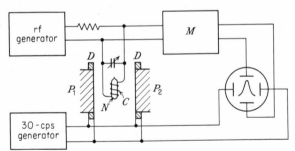

FIG. (3-1)32. Nuclear magnetic resonance arrangement for display of the absorption curve on the screen of a cathode-ray oscillograph: $P_1$, $P_2$, poles of magnet; $N$, sample; $C$, coil; $D$, auxiliary coils.

appear on the screen. The resonance frequency is read when the resonance maximum appears at the center of the screen, i.e., when the instantaneous value of the a-c field is zero. The resonance curve is sometimes complicated by transient phenomena.

For detailed descriptions of this and other experimental arrangements, see E. R. Andrew, "Nuclear Magnetic Resonance," chap. 3, Cambridge University

Press, London, 1955. For literature on the measurement of magnetic fields by magnetic nuclear resonance methods, see the same book, chaps. 4.9 and 4.10.

Energy can be absorbed from the electromagnetic rf field only if more nuclei are in the lower energy level. The continuous absorption of energy from the rf field has the effect of increasing the number of nuclei in the higher level. If the population of both levels is the same, the sample ceases to absorb energy from the rf field (the sample is "saturated"), and no absorption effect can be noted. In order to observe the nuclear-resonance effect, it is necessary, therefore, to work at very low rf field strength where no saturation effect occurs.

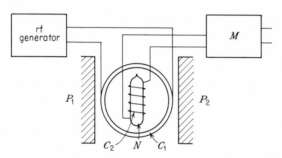

FIG. (3-1)33. Nuclear magnetic induction system: $P_1$, $P_2$, magnetic poles; $C_1$ and $C_2$, coils, axes perpendicular upon each other and upon permanent field $P_1$ and $P_2$; $N$, sample.

Energy from a nucleus in a higher level can be dissipated by the transfer of energy to the neighboring atoms in the lattice (and raising the lattice temperature). The nuclear-lattice interaction can be enhanced, and the observation of resonance absorption improved through the addition of a small amount of a paramagnetic ion, such as ferric nitrate, $Fe(NO_3)_3$, or, to a lesser extent, by using $H^1$ nuclei in highly viscous fluids.[1]

*b. Nuclear Magnetic Induction.* The arrangement is shown schematically in Fig. (3-1)33. It consists of a sample $N$ containing magnetically resonant nuclei ($H^1$ in water or hydrocarbon, $Li^7$ in a salt solution) in an rf energized coil $C_1$ between the poles of a magnet. A pickup coil is oriented at right angles to both the direction of the steady magnetic field and the axis of the rf coil. The change of magnetization induces a voltage in this coil which may be amplified and observed. The induced voltage is a maximum at the same fre-

[1] N. Bloembergen, E. M. Purcell, and R. V. Pound, *Phys. Rev.*, **73**, 679 (1948).

quency $f_0$ at which resonance absorption occurs. The physical process is similar to that of the absorption phenomenon.[1]

All nuclear-resonance methods require a good field homogeneity (gradient not more than several gauss per centimeter throughout the volume of the probe), since otherwise the resonance curve is too broad for convenient measurement. Small probes containing not more than 0.03 cm³ of water have been used successfully for measuring field gradients up to 300 to 400 gauss/cm. The value of the nuclear magnetic resonance method lies primarily in its high absolute accuracy. The experimental requirements are considerable.

### 3-15. INDIRECT SYSTEMS

The following systems do not furnish an electric signal in response to a magnetic input; they produce a mechanical force, a displacement, a thermal, or an optical output which is converted into an electric signal by additional means.

*a. Mechanical Methods.* FREELY SUSPENDED MAGNETS. In principle, any magnetometer, i.e., a system of freely suspended magnets,

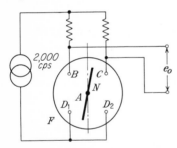

FIG. (3-1)34. Mechanoelectric magnetic-field transducer: $N$, magnet needle; $A$, pivot; $F$, capsule filled with semiconducting fluid; $B$, $C$, $D_1$, and $D_2$, contacts.

can be used as a magnetoelectric transducer, if means are provided to convert the angular displacements of the magnets into electric signals. A combination of this kind has been described by Gollian and Reilly[2] and is shown schematically in Fig. (3-1)34. A magnet needle $N$ is mounted in an enclosed capsule $F$ so as to turn freely in a horizontal plane around the pivot $A$. The capsule is filled with a semiconducting liquid (commercial ethyl alcohol); four contacts $B$, $C$, $D_1$, and $D_2$ are provided. The arrangement is connected in a bridge circuit. Any deviation of the needle from the symmetry position causes a bridge unbalance and produces a voltage $e_o$ at the output terminals. The output signal changes as an approximately parabolic function with the angular needle displacement, and varies from 0.3 to 1 volt for a variation of position from zero to 10°. The resistance between the electrodes $B$ and $D$ for symmetry position of the needle is in the order of 50,000 ohms.

[1] See Andrew, *op. cit.*, chap. 3.8, and F. Bloch, W. W. Hansen, and M. E. Packard, *Phys. Rev.*, **69**, 127 (1946), and *Phys. Rev.*, **70**, 474 (1946).

[2] S. E. Gollian and E. G. Reilly, *Rev. Sci. Instr.*, **22**, 753 (1951).

MOVING COIL MAGNETOMETER.  A moving coil, Fig. (3-1)35, with $n$ turns and the cross-sectional area $a$ ($= LW$) is suspended so that it can turn around an axis $yy$.  A current $I$ passes through the coil.  If the coil is placed in a magnetic field $H$ in the direction of the coil plane, a torque $\tau$ will tend to rotate the coil against the restoring

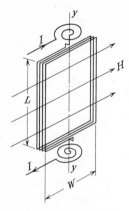

FIG. (3-1)35.  Moving-coil magnetometer.

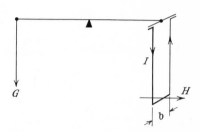

FIG. (3-1)36.  Wire-loop magnetometer.

force of the springs or of the suspension wires.  For a small angular displacement $\varphi$ the torque is

$$\tau = Ian\mu H = k\varphi$$

where $k$ is the spring constant.  If the current is regulated so that the same small angular displacement is always obtained ($\varphi = $ const), then the magnetic field can be found from

$$H = \frac{k\varphi}{an\mu I} = \frac{C}{I}$$

The apparatus constant $C$ can be found from calibration in known fields.[1]  The method is applicable for magnetic field strengths between 10 and 10,000 oersteds.  Traces of iron in the coil tend to introduce errors.

FORCE UPON A WIRE LOOP.  A rectangular wire loop of the form shown in Fig. (3-1)36 is suspended from the arm of a scale so that the lower (horizontal) bar of the loop reaches into a magnetic field of the strength $H$.  If a current $I$ passes through the loop, it will exercise a force downward or upward which is

$$F = bI\mu H$$

[1] H. S. Jones, *Rev. Sci. Instr.*, **5**. 211 (1934).

where $b$ is the length of the bar. The force can be balanced by placing a weight of $G$ grams on the other side of the scale.

The method is applicable for strong fields (order of 10,000 oersteds) and furnishes quite accurate results; the error is of the order of $\frac{1}{2}$ of 1 per cent.[1] The current is of the order of $\frac{1}{2}$ amp, the length $b$ about 0.5 cm. The sensitivity, i.e., the change of weight on the scale per change of magnetic field strength, is of the order of 0.1 mg/oersted.

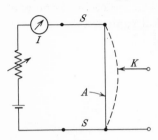

FIG. (3-1)37. Wire-loop magnetometer, alternate method.

A similar arrangement is shown in Fig. (3-1)37. A metal ribbon $A$ (of silver or aluminum) is held between two springs $S$. A current $I$ passes through the ribbon. In a homogeneous magnetic field $H$, perpendicular to the drawing surface, a force will act on each element of the conductor $A$ which is proportional to the product $I\mu H\ dl$. Under the influence of this force the wire will deflect. The current $I$ is varied until the wire just makes contact with the electrode $K$. The magnetic field strength can be found from

$$H = \frac{C}{I}$$

where $C$ is a constant found by experiment. The instrument is useful for field strengths above 10 oersteds.[2]

*b. Thermal Methods.* When a conductor is placed in an oscillating magnetic field, eddy currents are induced in the conductor and heat is generated. The resulting increase in temperature at a given frequency is a measure of the magnetic field strength.

The simplest embodiment of this principle is a mercury thermometer. The mercury bulb is placed in an rf magnetic field, and the temperature is read. The sensitivity of the arrangement can be modified by surrounding the thermometer with a metallic loop or with ferromagnetic materials in which additional heat is produced by hysteresis. Figure (3-1)38 shows calibration curves of a thermometer with a bulb of 40 mm length and 5.5 mm diameter. In the upper part of the curve, the sensitivity is relatively high, so that field variations of 0.1 per cent can be measured. The disturbance of the rf magnetic field by the presence of the metal may be considerable.

[1] A. Färber, *Ann. Physik*, **9**, 892 (1902).
[2] F. Schröter, *Z. Instrumentenk.*, **44**, 477 (1924).

Faster response (and presumably less disturbance of the field) can be obtained with a thermoelectric method. According to Laporte and Vasilesco,[1] a thermoelement junction in an evacuated vessel is placed in the magnetic rf field. The cold junction is outside the field and kept at constant temperature. The emf is measured with a galvanometer or microammeter. Figure (3-1)39 shows the thermocouple

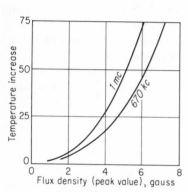

FIG. (3-1)38. Transfer characteristics for a thermal magnetometer in an rf magnetic field.

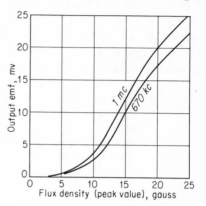

FIG. (3-1)39. Transfer characteristics, for two different frequencies, of a thermoelectric magnetometer in an rf field.

output as a function of the applied field strength for an iron-constantan thermoelement. In the linear part of the characteristic, the sensitivity is of the order of 1 mV/gauss.

   *c. Optical Methods.* FARADAY EFFECT AT OPTICAL FREQUENCIES. The principle of this transducer system is illustrated in Fig. (3-1)40.

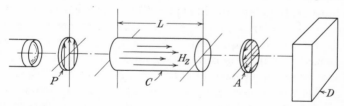

FIG. (3-1)40. Faraday-effect magnetic transducer, schematic diagram

A monochromatic parallel light beam passes an optical polarizer $P$, a vessel $C$ of length $L$ filled with a material which exhibits magneto-rotation, and an analyzer $A$ adjusted at an angle of 90° with respect to the polarizer, so that no light will fall upon the detector $D$. If a

[1] M. Laporte and V. Vasilesco, *J. physique radium,* **6,** 289 and 85 S (1935).

magnetic field $H_z$ is applied, as indicated by the arrow, the plane of polarization will be turned through an angle

$$\varphi = VL\mu H_z$$

$V$ is the Verdet constant of the substance in the vessel $C$. (More accurately, $V$ expresses the angle of rotation per centimeter of path length in the substance under the influence of a field strength of 1 oersted for the sodium D line of 589 m$\mu$.) Some values of Verdet constants are given in Table 14. The values vary with the temperature and with the wavelength of the light. The angle of rotation is usually determined by rotation of the analyzer.

TABLE 14. VERDET CONSTANTS AT ROOM TEMPERATURE, FOR Na D LINES
(In min/cm-oersted)

| | |
|---|---|
| $H_2O$ | $13.1 \times 10^{-3}$ |
| $SiO_2$ (quartz, perpendicular to axis) | 16.6 |
| $PbSiO_3$ (fused) | 77.9 |
| ZnS (zinc blende) | 225 |
| NaCl | 16 |
| Amber | −9.6 |
| Glass (0.500 heavy flint) | 60.8 |
| $CS_2$ | 43 |
| $CCl_4$ | 16 |

FARADAY EFFECT AT MICROWAVE FREQUENCIES. At optical frequencies the Verdet constant for most materials is quite small, but for ferrite materials at microwave frequency the Verdet constant reaches values which are about 100 times larger. The Verdet constant for some ferromagnetic ferrites, at 9,000 Mc, is of the order of 0.1°/(cm)(oersted).

In most instruments of this type, the analyzer is set at 90° with respect to the polarizer, so that at zero field strength no radiation reaches the detecting apparatus. As the field is increased, the plane of polarized radiation is rotated by an angle $\varphi$, and the radiation intensity, which is proportional to sin $\varphi$, is measured. Rectangular waveguides which have only one dominant mode of excitation can be used for polarization. Two waveguide sections turned at 90° serve as effective polarizer and analyzer. The microwave signal can be detected with a probe and rectifier. The output may be amplified and read on a vacuum-tube voltmeter. The detected electric vector is

$$E_o = E_i \sin \varphi$$
$$= E_i \sin Vl\mu_a H_z \tag{1}$$

where $E_i$ is the amplitude of the incident electric vector, $V$ the Verdet constant, $l$ the length of the ferrite body, and $H_z$ the magnetic field strength in the lengthwise direction of the ferrite. The value of $\mu_a$, the apparent permeability of the ferrite, is a function of $\mu$, the true permeability, and the physical size of the specimen; $\mu_a$ increases linearly with $l/d$ (ratio of length to diameter) and approaches $\mu$ as $l/d$ approaches infinity. It is desirable to select a ferrite with high $\mu$.

If the ferrite transmits with little loss and the rotation has been very slight, the analyzer will reflect most of the power back to the source through the ferrite. To prevent this reflection and still detect a signal, an absorber must be inserted into the waveguide after the ferrite, which will absorb any component polarized in the same direction as the incident microwaves. If the fer-

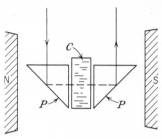

FIG. (3-1)41. Optical magnetic transducer based upon the orienting effect of magnetic fields upon colloidal particles in a suspension.

rite has a large absorption coefficient, this is not necessary, but in this case the ferrite lowers the intensity, and Eq. (1) must be multiplied by the $e^{-l\alpha}$ where $\alpha$ is the absorption coefficient.

The method is applicable to the measurement of magnetic fields in a range from $10^{-5}$ to 10 oersteds. In the range above 10 oersteds the ferrite tends to become saturated, and the output characteristics become nonlinear. The sensitivity is very high; flux-density variations of the order of $10^{-5}$ gauss and rotations of $10^{-4}$ deg have been detected. The field should be uniform throughout the volume of the ferrite. The method is limited by noise introduced by the ferrite, by inhomogeneity of the ferrite, and by temperature variations.[1]

LIGHT ABSORPTION IN COLLOIDAL SUSPENSIONS. Figure (3-1)41 shows a schematic diagram of an arrangement in which a variation of light absorption in colloidal suspensions takes place under the influence of a magnetic field.[2] A vessel $C$ is placed in the magnetic field between the poles of a magnet. A light beam passes through the two prisms $P$ and through the cell in a direction parallel to the field. The cell is filled with a colloidal suspension of particles that are opaque, anisotropically diamagnetic or paramagnetic, and nonspherical, such as colloidal graphite (diamagnetic) in water. The absorption of light in this suspension depends upon the projected

[1] P. J. Allen, *Proc. IRE*, **41**, 100 (1953), and *Rev. Sci. Instr.*, **25**, 394 (1954).

[2] F. D. Stott and A. von Engel, *Nature*, **161**, 728 (1948).

area of the graphite particles. A magnetic field has the effect of orienting the anisotropically diamagnetic particles; the thermal motion of the particles tends to reestablish random orientation. An exact theory of the process is given by Stott.[1] The optical transmission along the magnetic field increases by about 30 per cent for a field variation from about 100 to 3,000 oersteds. Similar experiments were carried out by Mueller and Shamos[2] with $Fe_2O_3$ in mineral oil and $CCl_4$ in fields below 10 gauss. At higher field strength saturation sets in. The orientation time for the particles in a magnetic field is relatively short, while the decay time is in the order of 0.1 to 1 sec for small graphite particles and even longer for $Fe_2O_3$ particles in oil.

[1] F. D. Stott, *Proc. Phys. Soc.* (*London*), (**B**) **62**, 418 (1949).
[2] H. Mueller and M. A. Shamos, *Phys. Rev.*, **61**, 631 (1942).

# 4

## Electrical Transducers

Electrical transducers convert certain electrical quantities into electric signals. An electric signal, as defined in the introduction, is either a voltage, a current, an impedance, or a variation or a time function of these quantities. The input quantity applied to the transducer is of an electrical nature but does not have the character of an electric signal; it may be a free electric charge, such as free electrons or ions (4-1), or a space charge (4-2), or it may be a space potential (4-3), a surface charge or surface potential (4-4), or an electric field strength (4-5).

### 4-1. Transducers Responding to Free Charges

The two most frequently used transducer systems for the measurement of free electrons and ions with a kinetic energy larger than thermal energy are the charged plate (4-11a) and the Faraday cage (4-11b). Positive and negative charges of thermal energy in atmospheric air can be measured with the Gerdien ion counter (4-12a). A system for the measurement of the excess charges in air (positive minus negative charges) is described under 4-12b. A method for the measurement of moving ion clouds is described in 1-68.

Electrons, positive ions, and alpha particles can also be measured with electron multipliers, similar to the photomultipliers described in 5-11c. The efficiency of the electron multiplier for incident electrons of an energy of 500 eV is almost 100 per cent, but decreases for lower as well as for higher energy.[1]

[1] J. S. Allen, *Rev. Sci. Instr.*, **18**, 739 (1947), and Z. Bay, *Rev. Sci. Instr.*, **12**, 127 (1941); literature references in both papers.

Electrons and ions which possess sufficient kinetic energy to ionize can also be measured with ionization chambers (5-21$a$), Geiger counters (5-21$b$), and scintillation counters, as well as with the cloud chamber, with a calorimeter, and by their photographic action or by fluorescence.

## 4-11. Electrode Systems

*a. Charged-plate Electrode.* The simplest receiver for electric charges consists of an insulated metal plate arranged in the beam of the charged particles, as shown for negatively charged particles in Figs. (4-1)1 and (4-1)2. The incident charged particles discharge the

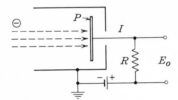

Fig. (4-1)1. Transducer for electric charges; charged plate.

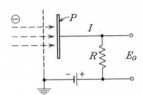

Fig. (4-1)2. Charged plate with grounded screen.

plate $P$ and form a current $I$ through the (large) resistance $R$. To avoid the escape of secondary electrons caused by the impact of primary electrons and ions upon the plate or of reflected or back-scattered primary electrons, the plate should be charged to a positive potential with respect to the walls of the surrounding vessel, as in Fig. (4-1)1, or with respect to a grounded screen located in front of the plate, as in Fig. (4-1)2. The potential of the plate should be at least that corresponding to the electron-volt velocity of the incident charges. A charged electrode for the measurement of local electron intensities in an electron microscope is described by Bahr, Carlsson, and Lomakka.[1]

A modification of the charged plate is the so-called paraffin capacitor, a metal plate covered with a thin layer of paraffin over which a metal layer is deposited; the incident electrons induce charges in the plate.[2]

*b. Faraday Cage or Cup.* The Faraday cage, shown in Fig. (4-1)3, consists of a hollow metallic body $C$. A beam of charged particles

[1] G. F. Bahr, L. Carlsson, and G. Lomakka, *Rev. Sci. Instr.*, **27**, 749 (1956)

[2] P. Lenard, *Ann. Physik*, **64**, 288 (1898); A. Becker, *Ann. Physik*, **13**, 394 1904. R. Thaller, *Physikal. Z.*, **29**, 841 (1928).

enters the cage through a small opening. Reflected primary charges and secondary electrons traverse the inside of the cage at different angles, but their chances to leave the cage are very small, so that practically all incident charges are measured. A diaphragm $D$ in front of the opening prevents incident charges from reaching the outside of the cage where they could produce secondary electrons. The diaphragm has either the form of a disk in front of the opening side of the cage or it surrounds the cage and serves as an electrostatic shield.

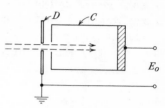

Fig. (4-1)3. Faraday cage.

The few electrons leaving the cage have in general a velocity of the order of 100 eV and can be driven back into the cage by the application of a potential between the cage and the shield or by a magnetic field. For high-energy particles, the bottom of the cage should be made from a material with low secondary emission, such as beryllium or carbon. Most of the secondary electrons emitted from such surfaces have an energy of only a few hundred electron-volts, even if the primary energy is about 400 keV. The use of such materials in the cup tends to reduce the number of double secondaries from the walls of the cup.[1]

Errors are likely to arise from the collection of electrons and ions in the vicinity of the cup. Enclosing the cup in an evacuated envelope reduces this effect. Further errors may arise if high-energy particles penetrate the cup; the influence may be reduced by using a thick layer of carbon and lead at the bottom of the cup. A Faraday-cup monitor for electron beams of an energy ranging from 4 to 300 MeV with an error of less than 0.5 per cent is described by Brown and Tautfest.[2]

The potential to which the Faraday cage is charged is proportional to the charge and the number of the incident particles and is independent of their kinetic energy. For large incident beam currents, the current from the cage to ground (or to any other reference electrode) can be measured with a galvanometer or by measuring the voltage drop across a large resistor. If the beam current is small, it is more advantageous to accumulate a charge on the cage during the time $t$ and measure the cage potential $E$. This potential should be small to reduce leakage currents. If $E$ is small compared with the

[1] J. G. Trump and R. J. Van de Graaff, *Phys. Rev.*, **75**, 44 (1949).
[2] K. L. Brown and G. W. Tautfest, *Rev. Sci. Instr.*, **27**, 696 (1956).

beam energy expressed in electron-volts, then the beam current can be found from

$$I = \frac{CE}{t}$$

where $C$ is the capacitance to ground of the cage.

The charged plate and the Faraday cage permit the measurement of a charge as small as $10^3$ to $10^4$ electrons.[1] Large errors are likely to occur from inadequate insulation.

The charged plate and the Faraday cage can be used to measure the velocity distribution in a beam of charged particles. For this purpose an opposing potential is applied to the cage, and the beam current is measured as a function of this potential. The arrangement is shown in Fig. (4-1)4a for electron beams. With increasing voltage $-E$ an increasing number of electrons are unable to enter the cage and the current decreases as shown in Fig. (4-1)4b. The differentiation of this curve furnishes the velocity distribution curve shown in Fig. (4-1)4c.

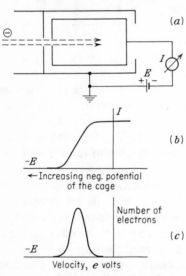

FIG. (4-1)4. Application of the Faraday cage for the determination of electron velocity distribution: (a) circuit diagram; (b) voltage-current characteristic; (c) differentiated voltage-current characteristic.

**4-12. GAS-FLOW SYSTEMS**

*a. Gerdien Ion Collector.* This system is used primarily for the determination of the number and the concentration of ions in gases (air) but can also be used for the measurement of ion mobility and air conductivity. The system is illustrated schematically in Fig. (4-1)5 and consists of a cylindrical capacitor, the external cylinder being grounded and the internal electrode $C$ charged to a voltage $E$. Air containing ions is drawn through the capacitor; the ions are attracted toward the electrode $C$ and cause in the external circuit a current which is proportional to $n$, the concentration of ions, i.e., the number of ions per cubic meter, to their charge $q$ and to the

[1] O. Klemperer, "Einführung in die Elektronik," p. 30, J. Springer Verlag, Berlin, 1933.

volume flow velocity $M$ in cubic meters per second. If the voltage $E$ is sufficiently high so that all ions are deposited on the electrodes, the current is

$$I = nqM \tag{1}$$

The voltage-current characteristic of the counter is shown in Fig. (4-1)6. The law expressed by Eq. (1) is valid only in the saturation region, i.e., at voltages above $E_s$. In this region, the system can be used for the determination of $nq$, the concentration of charges, or, if $q$ is known, the concentration of ions ($n^+$ or $n^-$). Positive and negative charges are counted separately by applying either a negative or a

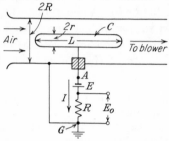

FIG. (4-1)5. Gerdien ion collector.

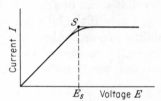

FIG. (4-1)6. Voltage-current characteristics of the ion collector, used for the determination of the ion mobility.

positive potential to the insulated electrode $C$. The ion currents are frequently too small to be measured by the voltage-drop method indicated in Fig. (4-1)5; higher sensitivity is obtained by electrometer methods. The electrometer is inserted between the points $A$ and $G$; it is charged initially and should not be allowed to discharge to a value below $E_s$.

In the region below $E_s$ not all ions are drawn to the electrode; the current then depends upon the ion mobility $k$ (i.e., the velocity of the ion, in meters/second, if exposed to a field of 1 volt/m). In this region the current is

$$I' = mqn \frac{k}{k_g} \tag{2}$$

where $k_g$ is the "threshold mobility," i.e., the mobility of those ions which are just driven to the electrodes. Ions with a mobility smaller than $k_g$ are not deposited at the electrodes. The threshold mobility is a constant of the apparatus, it depends upon the counter geometry and the applied voltage and is, for the parallel-plate capacitor with homogeneous flow throughout the volume between the plates,

$$k_g = \frac{\epsilon_0 M d}{aE} \tag{3a}$$

($d$, distance between plates; $a$, area of the plates) and for the cylinder capacitor

$$k_g = \frac{2\epsilon_0 M(\ln R/r)}{LE} \tag{3b}$$

For $R$, $r$, and $L$, see Fig. (4-1)5.

In the region below $E_s$ the Gerdien capacitor can thus be used to measure air conductivity, defined by $nk$, the product of ion concentration and ion mobility.

The ion mobility $k$ can be determined by finding experimentally point $S$ in Fig. (4-1)6; at this point the ion mobility $k$ is equal to $k_g$, so that $k$ can be found from Eqs. (3a or b). A sharp bend at $S$ is present only if the ions in the gas stream have the same charge and mobility. In any other case the characteristic shows several bends from which, in simple cases, the presence of the components (ion spectrum) can be found.

For summarizing reviews and references, see H. Israel, *Arch. tech. Messen*, V 656–2, August, 1953, and V 656–3, June, 1955; also O. W. Torreson, in J. A. Fleming (ed.), "Physics of the Earth," Chap. 5, McGraw-Hill Book Company, Inc., New York, 1939. An airborne "ion filter" is described by R. Gunn in H. R. Byers (ed.), "Thunderstorm Electricity," Chap. 8, pp. 199ff., The University of Chicago Press, Chicago, 1953.

*b. Excess of Ions of One Polarity.* A metal cylinder $A$ packed with fine-grade steel wool is held, electrically insulated, within a grounded-metal shield $B$, as shown in Fig. (4-1)7. The cylinder $A$ is initially discharged. Air containing ions of either polarity is drawn through the steel wool; the ions are discharged in contact with the steel wool and give their charge to the cylinder $A$. If the concentration of positive and negative charges is equal, the cylinder $A$ will remain uncharged. However, if more ions of one polarity are present, the cylinder will become charged to that polarity, and its voltage will be

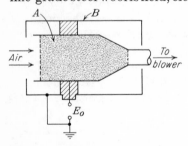

FIG. (4-1)7. Excess-ion collector: $B$, grounded container; $A$, insulated electrode.

$$E_o = \frac{n' M t}{C}$$

where $n'$ is the concentration of the excess charge (i.e., the number per cubic meter of charge of one polarity minus that of the other polarity), $M$ the gas-volume flow velocity in cubic meters per second, $t$ the collection time in seconds, and $C$ the capacitance to ground (shield) of the cylinder $A$.

The method has been used in atmospheric-electricity studies to determine the space charge, i.e., the local concentration of excess charge in the air; excess-charge concentrations from $3.3 \times 10^{-12}$ to $160 \times 10^{-12}$ coulomb/m³ have been measured.[1]

## 4-2. Space Charges

Space charge is the local concentration of positive ions $n^+$ or of electrons and negative ions $n^-$ in a medium.

The classical transducer for the space-charge determination in highly ionized gases is the Langmuir probe (4-21).

The space charge in atmospheric air is defined as the local excess concentration of ions of one polarity. It can be determined by measuring the concentration of ions of either polarity, $n^+$ and $n^-$, with the Gerdien counter (4-12a) and forming the difference $n^+ - n^-$. Alternatively, the Brown excess-ion counter (4-12b) can be used; its output is directly proportional to the space charge as defined above.

In cases where the space is well localized and the geometry of the electric field is well defined, the space charge can be determined from electric-field measurements (4-5).

### 4-21. LANGMUIR PROBES

In highly ionized gases, with a concentration of ions larger than $10^8$ per cubic centimeter, the space charges and the space potential can be found from the voltage-current characteristics of a probe inserted in the ionized gas (plasma).

The probe arrangement is shown in Fig. (4-2)1. Figure (4-2)2a illustrates the usual voltage-current characteristic. If the probe potential $E_p$ is highly negative with respect to the plasma potential (point $A$), electrons will not be able to reach the probe; only a constant positive ion current $i^+$, independent of the probe potential, is noted.

If the probe potential is made less negative (point $B$), some electrons will reach the probe, and with further reduction of the negative potential, the electron current (obtained by deducting the ion current $i^+$ from the total current $i$) will increase exponentially. By plotting the logarithm of the electron current density, $j = (i/a)$, to the probe

---

[1] J. G. Brown, *Terrestrial Magnetism and Atm. Elec.*, **35**, 1 (1930), and **38**, 161 (1933).

versus the probe voltage $E_p$, Fig. (4-2)2b, a straight line $\alpha$ is obtained. The slope of this line depends upon the electron temperature $t^-$ (assuming a Maxwell velocity distribution) so that $t^-$ can be found from

$$t^- = -\frac{e}{k}\frac{dE_p}{d\log j^-} \tag{1}$$

If the probe potential is made less negative, the electron current will be equal to the ion current, so that the total current is zero

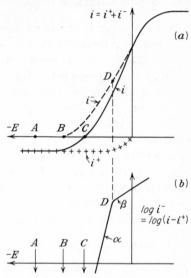

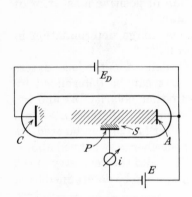

FIG. (4-2)1. Langmuir probe: $C$, cathode; $A$, anode of discharge tube; $E_D$, discharge voltage source; $P$, probe; probe potential adjusted by varying voltage source $E$; $S$, ion sheath formed in front of negatively biased probe.

FIG. (4-2)2. Voltage-current characteristic of the Langmuir probe.

(point $C$). With further reduction of the negative probe potential, the slope $\beta$ of the log $i$ versus $E_p$ characteristic changes. The bend of the characteristic occurs when the probe potential is equal to the space potential $E_s$. At this point, provided no collision occurs in the sheath, the electron current is approximately

$$i^- = aen^-\frac{kt^-}{2\pi m^-} \tag{2}$$

where $e$ is the charge, $m$ the mass of the electron, and $k$ the Boltzmann constant. The quantity $a$ is the area on the plasma side of the sheath covering the probe; with a plane probe, $a$ is approximately equal to the surface area of the probe. The electron current $i^-$ can be computed from Eq. (2). The positive-ion concentration $n^+$ is about equal to the electron concentration $n^-$. Errors are likely to arise from surface contamination of the probe.

For theoretical consideration and alterations of the theory due to probe geometry (cylindrical and spherical probes), see A. Von Engel, "Ionized Gases," pp. 262ff., Oxford University Press, London, 1955, and G. Francis, "The Glow Discharge at Low Pressures," in S. Flügge (ed.), "Encyclopedia of Physics," vol. 22, pp. 62ff., Springer Verlag, Berlin, 1956.

## 4-3. Space Potential

A space potential is the potential at a given point in an electric field. It is usually expressed as the potential difference between this point and a fixed electrode (e.g., ground).

The space potential in an electric gas discharge with a high ion concentration can be measured with the Langmuir probe (4-2). For the measurement of space potentials in the earth's atmosphere, special probes ("collectors") have been developed; they are described in 4-31. The space potentials in electron optical arrangements can be measured with a system described in 4-32. The application of this system is limited to special cases; more frequently the space potential is found from model experiments in the electrolytic trough (4-33).

### 4-31. Potential Probes, Collectors

A space-potential probe as used for atmospheric-electricity studies generally consists of a metallic electrode $S$, well insulated by a rod $A$

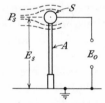

FIG. (4-3)1. Potential probe, collector: $S$, metal electrode; $A$, insulated rod.

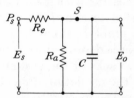

FIG. (4-3)2. Equivalent circuit of the potential probe illustrated in Fig. (4-3)1.

from ground (or from the reference electrode), as shown in Fig. (4-3)1. The electrode $S$ should be small compared to the dimensions of the field, so that the field is practically homogeneous in the vicinity of the probe. When exposed to an electric field, the electrode $S$ will gradually assume the local potential $E_s$ by exchange of charges with the environment; i.e., the ions and electrons in the atmosphere will cause a charge or discharge of the probe until its potential is equal to the space potential at the place of the probe.

This exchange of charges between the probe and its surroundings is a slow process and may lead to erroneous readings if the field and the space potential vary with time or if the amount of charges carried from or to the electrode $S$ becomes comparable to that leaking off through the insulator. The situation is evident from Fig. (4-3)2, where the charge-discharge path between the probe $S$ and the surrounding potential $P_s$ is indicated by an equivalent resistor $R_e$, the capacitance between the insulated probe and ground by $C$, and the insulation resistance by $R_a$. Only for $R_a \gg R_e$ is the voltage $E_o$ across the capacitor equal to the space potential $E_s$. A number of systems have been developed which permit a faster exchange of charges between the probe $S$ and its environment and lead to a reduction of the equivalent resistance $R_e$ to a value between $10^9$ and $10^{12}$ ohms.[1]

A reduction of the magnitude of $R_e$ has the further advantage of reducing the time constant of the probe system, which, for $R_a \gg R_e$, is proportional to $R_e C$. Since $C$ (including the electrometer connected to it) is of the order of 100 $\mu\mu$F, the time constant of a collector is between 0.1 and 100 sec.

WATER-DROP COLLECTOR. A container with water is connected with the probe electrode. The water is either permitted to drip from the container or is forced by compressed air through one or several nozzles, so that it leaves the container in the form of a spray. Each droplet carries with it a certain amount of charge (the magnitude and polarity determined by electrostatic induction), until the potential of the probe is equal to the space potential. An airborne water collector is described by Gunn.[2]

Drawbacks of the system are the need for replenishment of the water, the insulation difficulty of the water container if located indoors and connected to the outdoor probe, or the danger of freezing.

IONIZATION COLLECTORS, POINT-DISCHARGE COLLECTOR. The collector is connected with a fine point. The electric field strength at the point is high so that ionization occurs. The discharge current from the point to the surrounding field is

$$i = a(F^2 - M^2)$$

[1] H. Israel, *Arch. tech. Messen*, V 656–3, June, 1953.

[2] R. Gunn in H. R. Byers (ed.), "Thunderstorm Electricity," Chap. 8, pp. 199ff., The University of Chicago Press, Chicago, 1953.

where $F$ is the field strength prevailing at the point under operating conditions and $M$ the minimum field strength of the point at which a discharge occurs. Both $M$ and the proportionality constant $a$ depend upon the radius of curvature of the point.[1] The indication of the point collector is influenced by the wind velocity.[2]

The system is primarily useful for the measurement of relatively large potentials and field strengths (atmospheric studies of distant thunderstorms).

FLAME COLLECTOR. The collector contains a candle or a kerosene burner, the escaping flame gases contain ions which carry the charge off the probe. The system is simple and useful for orientating experiments, but it is not very stable and is rarely used nowadays.

RADIOACTIVE COLLECTOR. The probe contains a radioactive material which ionizes the air in the vicinity of the collector. Materials emitting $\alpha$ rays are preferred because the effect of $\alpha$ rays is limited to the immediate vicinity of the probe, so that the field at greater distance from the point remains undisturbed. The radioactive material is generally deposited on small metal plates or foils and protected by a thin lacquer or enamel layer. The collector is most suitable for the measurement of small, relatively steady potentials ("fair-weather potential gradient").

### 4-32. ELECTRON OPTICAL SYSTEM

A simple system for the measurement of space potentials in electron tubes (concentric diodes) is described by Kenyon.[3] The inner side of the anode is coated with a fluorescent layer, which can be observed. A probe consisting of a fine tungsten wire is introduced in the space between cathode and anode parallel to the tube axis, i.e., in an equipotential surface, to avoid field disturbances. The probe is connected to a potential between those of the anode and the cathode. If the probe potential is equal to the space potential, the probe will not alter the flow of electrons; if its potential is different from the space potential, it will cause a shadow or a bright line on the fluorescent screen.

The accuracy of the method is about 1 per cent for measurements in the space between cathode and about three-quarters of the way toward the anode, but diminishes as the probe approaches the anode.

---

[1] F. J. W. Whipple and F. J. Scrase, *Geophys. Mem.* (*London*), no. 68.

[2] J. A. Chalmers and W. W. Mapleson, *Repts. Progr. in Physics,* **17,** 101 (1954).

[3] D. E. Kenyon, *Rev. Sci. Instr.,* **11,** 308 (1940).

## 4-33. ELECTROLYTIC TROUGH

Space-potential distributions in electrostatic fields can be found from model experiments in the electrolytic trough. For similar electrode configurations, the distribution of the stationary field in a homogeneous conductor (electrolyte) is analogous to that of an electrostatic field in a homogeneous dielectricum.

The electrolytic trough[1] consists of a shallow insulating container, as shown in Fig (4-3)3, filled with an electrolyte. Metal electrodes $A$ and $B$ are inserted so that the field configuration of this model is analogous to that of the simulated system. The electrodes are connected to an a-c source $S$ (to avoid polarization). If a conductive stylus $P$ is inserted into the electrolyte, its potential can be found by comparison with that along the calibrated resistor $R$. $M$ is a zero indicating system (e.g., a headphone). By setting the movable contact $C$ at fixed points on the resistor $R$ and moving the stylus $P$ so that the meter $M$ does not indicate any potential difference, equipotential lines can be drawn. The electric-force lines are the trajectories to the equipotential lines.

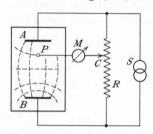

FIG. (4-3)3. Electrolytic trough: $A$ and $B$, electrodes immersed in electrolyte; $P$, stylus; $C$, variable contact on resistive voltage divider $R$; $M$, zero indicator.

Conductive paper (Teledeltos Paper, Western Electric Company) can be used in two-dimensional systems, instead of the electrolytic trough, but it does not furnish as accurate results as an electrolyte because of local variation of the conductivity of the paper.

The electrolytic trough chosen should be as large as is practical so that errors resulting from the influence of the walls upon the field distribution are avoided. Polarization can be avoided by making all the electrodes from the same material and as large as possible. The resistivity of the electrolyte should be reasonably high. The stylus usually consists of a thin wire; good results have been obtained with a stylus consisting of a tungsten wire of 0.001 in. diameter fused into the end of a glass capillary. Unavoidable capacitances between electrodes and probes cause an incomplete zero reading. The effect can be reduced by capacitive balance, i.e., inserting of variable capacitors between $A$ and $P$, and $B$ and $P$.

For review, see R. Strigel, *Arch. tech. Messen*, V 312–1, February, 1943.

[1] C. L. Fortescue and S. W. Farnsworth, *Proc. AIEE*, **32**, 757 (1913).

## 4-4. Surface Charges and Surface Potentials

### 4-41. WILSON PLATE

A simple system for the measurement of surface charges which are induced by relatively large electric fields is shown in Fig. (4-4)1. The system[1] uses an insulated electrode $P$ which, for practical measurements of atmospheric electricity, is flush with the ground but insulated from it. In other applications it may be surrounded by a guard ring.

For the measurement in d-c fields, the plate is first shielded by a grounded cover $S$ and momentarily discharged. The shield $S$ is then

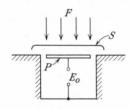

FIG. (4-4)1. Wilson plate as a transducer for surface charges.

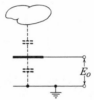

FIG. (4-4)2. Equivalent circuit of the Wilson plate.

withdrawn and the plate $P$ is exposed to the electric field. The field causes a surface-charge density

$$\sigma = \frac{CE_o}{a}$$

where $C$ is the capacitance of the plate to ground (including that of the electrometer connected to the output terminals), $E_o$ the voltage, and $a$ the surface area.

For the measurement of time-varying surface charges or of time-varying electric field causing surface charges, a simple electrode, such as the uncovered Wilson plate, is usually sufficient ("field variometer"). The equivalent circuit for a collector electrode or a Wilson plate at variational field strength is shown in Fig. (4-4)2.

A number of mechanical systems ("field mills") have been described which, in effect, cause a continuously repeating shielding and exposure of the plate; see J. A. Chalmers, "Atmospheric Electricity," chap. IV, par. 73, p. 63, Oxford University Press, New York, 1949; R. Gunn, *Phys. Rev.*, **71**, 181 (1947); see also N. Russelvedt, *Jahrb. Met. Inst. (Oslo)*, 1925 (1926).

[1] C. T. R. Wilson, *Phil. Mag.*, **17**, 634 (1909), *Phil. Trans. Roy. Soc. (London)*, **(A) 92**, 555 (1916), and **(A) 221**, 73 (1920).

## 4-42. VIBRATING CAPACITOR

The surface potential or contact potential[1] between two electrodes varies with the chemical and physical structure of the electrode surfaces; the observation of the variations of the potential of an electrode versus that of a constant-reference electrode can furnish information of a chemical or physical process taking place at the surface. Surface layers are sometimes vulnerable, so that measurements of the contact potential must be carried out without touching the surface.

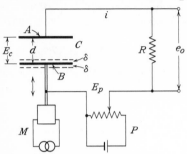

FIG. (4-4)3. Vibrating capacitor: $C$, capacitor; $A$, fixed, and $B$, vibrating electrode driven by motor $M$; excursion of the vibrating electrode is $2\delta$.

A system suitable to measure such potential differences was first described by Zisman[2] and is illustrated in Fig. (4-4)3. It consists of two electrodes $A$ and $B$ which together form a capacitor $C_0$. The surface of one electrode may be altered by applying the material to be investigated to it.

If $E_c$ is the potential difference between the electrode surfaces, $Q = C_0 E_c$ is the charge of the capacitor. Any change of capacitance will cause a current (neglecting $R$) of

$$i = \frac{dQ}{dt} = E_c \frac{dC_0}{dt} + C_0 \frac{dE_c}{dt} \tag{1}$$

Under steady-state condition ($E_c = $ constant), the second term in Eq. (1) vanishes.

A driving mechanism $M$ causes one electrode to oscillate with a frequency $f$ (angular velocity $\omega = 2\pi f$) by an amount $\pm\delta$ about its middle position. If $\delta$ is small compared to the average distance $d$, the capacitance can be expressed by

$$C = C_0\left(1 - \frac{\delta}{d}\sin \omega t\right)$$

and the current is

$$i = -E_c \omega C_0 \frac{\delta}{d}\cos \omega t$$

---

[1] For nomenclature see A. A. Frost, *Rev. Sci. Instr.*, **17**, 266 (1946).

[2] W. A. Zisman, *Rev. Sci. Instr.*, **3**, 367 (1932).

Usually a voltage rather than a current output is required and can be obtained by the insertion of a resistor $R$, as shown in Fig. (4-4)3. If $R \ll 1/C$, then the output voltage is

$$e_o = -E_c R \omega C_0 \frac{\delta}{d} \cos \omega t$$

The method can be transformed into a null method by inserting in the circuit an adjustable voltage $E_p$ derived from a potentiometer. If $E_p$ is equal to $E_c$ and of opposite polarity, the output voltage $e_o$ disappears. Changes of potential of the order of 10 $\mu$V can be measured.

The method is very sensitive. The presence of polar vapors (such as ethyl alcohol) in the laboratory air of a concentration by weight of 1 part in $10^7$ can change the surface potential of a stainless-steel plate by voltages in the millivolt range.[1] A gold-plated reference electrode covered with a layer of calcium palmitate about 20 molecules thick is recommended.

For further references, see A. A. Frost, *loc. cit.*; S. Rosenfield and W. M. Hoskins, *Rev. Sci. Instr.*, **16**, 343 (1945); W. E. Meyerhof and P. H. Miller, Jr., using a single forward movement of one electrode, *Rev. Sci. Instr.*, **17**, 15 (1946); H. H. Kolm, apparatus with rotary movement, *Rev. Sci. Instr.*, **27**, 1046 (1956); H. G. Yamins and W. A. Zisman, *J. Chem. Phys.*, **1**, 656 (1933).

### 4-43. SURFACE CHARGES ON INSULATORS

The (usually irregularly distributed) electrostatic charges on the surface of an insulator can be detected and measured by the use of a moving electrode. An instrument of this type is described by Devins and Reynolds.[2] It consists essentially of a fine wire, its end surface (area, 0.05 mm$^2$) scanning in a rotatory motion the surface to be examined at a distance from the surface of about 1 mm. A charge on the insulator surface changes the potential of the wire by electrostatic induction. The resolution is of the order of 1 mm.

## 4-5. Electric Fields

In general, the local electric field strength $F = dE/ds$ in a d-c field cannot be converted directly into an electric signal. Indirect-acting systems furnishing an electric output in response to electric d-c field strengths are the following:

[1] G. Phillips, *J. Sci. Instr.*, **28**, 342 (1951).
[2] J. C. Devins and S. I. Reynolds, *Rev. Sci. Instr.*, **28**, 11 (1957).

### 4-51. INDIRECT SYSTEMS

*a. Space-potential Systems.* Two probe electrodes (collectors) $P_1$ and $P_2$ of the type described in 4-31 are brought into the electric field, as shown in Fig. (4-5)1. The distance between the probes in the direction of the field is $\Delta s$; the voltage measured between the probes is $\Delta E$. The field strength can be approximated from

$$F = \frac{\Delta E}{\Delta s}$$

Frequently only one probe is used, and the potential between the probe and one field-terminating electrode (usually ground) is

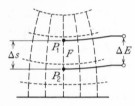

measured. In variational (a-c) electric fields, the space-potential measurement is, in principle, simpler than in d-c fields, since the field variation causes, by electrostatic induction, the generation of an a-c potential between the electrodes, as in Fig. (4-4)2. The connecting leads should be in the direction of the equipotential lines to avoid field deformation.

FIG. (4-5)1. Electric field $F$ with two probes, $P_1$ and $P_2$.

*b. Surface-charge Systems.* If an insulated electrode is exposed to an electric field, such as the Wilson plate illustrated in Fig. (4-4)1, the field will cause a surface charge of the electrode by electrostatic induction. The field strength $F$ in the immediate vicinity of the electrode is related to the surface-charge density $\sigma$ by

$$F = \frac{\sigma}{\epsilon_0} = \frac{Q}{\epsilon_0 a} = \frac{CE_o}{\epsilon_0 a}$$

where $\epsilon_0$ is the dielectric constant, $Q$ the charge of the electrode, $a$ the electrode surface, $C$ the capacitance of the electrode against ground, and $E_o$ the voltage measured against ground.

Any one of the surface-charge systems described in 4-4 can be used. The Wilson plate is frequently used in atmospheric-electricity studies, primarily for measurement during violent thunderstorms.

### 4-52. METHOD BASED UPON MOVING CHARGES (ELECTRON-BEAM PROBE)

The following method is primarily applicable to the measurement of electric fields in evacuated systems or such filled with gases under reduced pressure (glow discharges). As shown in Fig. (4-5)2, an electron beam $B$ emitted and focused by an electron gun $G$ passes through the electric field in a gas discharge that takes place between the two electrodes $C$ and $A$. The electric field causes a deflection of the electron beam which can be observed on the fluorescent screen $F$.

The method is very sensitive. Warren,[1] who has combined this system with a feedback arrangement that compensates the beam deflection by an auxiliary magnetic field, has measured fields from 1,000 down to 0.2 volts/cm.

The gas in the tube tends to cause a diffusion of the electron beam. With a gas discharge tube of a diameter of 10 cm and a beam velocity of about 20 kV, the application of the method is limited to pressures below about 1 mm Hg.

For further literature references, see the paper by Warren, *ibid*.

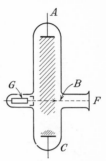

FIG. (4-5)2. Electron-beam system for the measurement of electric fields: $A$ and $C$, electrodes between which an electric gas discharge passes; $G$, electron gun; $B$, electron beam; $F$, fluorescent screen.

### 4-53. METHOD BASED ON LUMINOUS-GAS DISCHARGE IN RF FIELDS

A system for the measurement of high field strengths (of the order of several hundred volts per centimeter) in rf fields has been described by Lion.[2] The system is shown schematically in Fig. (4-5)3. A spherical tube $B$ containing gas under reduced pressure is brought into the electric rf field. If the field strength is sufficiently high, an electrodeless glow discharge will arise in the tube. The tube is surrounded by an insulating, light-tight shield $S$ which may be constructed as a light guide. The light emitted from the tube falls upon a photoelectric cell $P$ which is outside the field and which is connected to an amplifier and a meter.

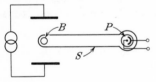

FIG. (4-5)3. Electric-field transducer for measurement in strong rf fields: $B$, electrodeless discharge tube; $S$, tubular light shield; $P$, photoelectric receiver.

Calibration curves (light emitted by the tube $B$ versus field strength) are shown in Fig. (4-5)4. The transfer characteristics are linear, but their slope varies with the frequency of the rf field.

The system avoids metallic parts in the field, the glow tube does not disturb the field appreciably, and the power consumption of the tube is, in general, a fraction of a watt.

[1] R. W. Warren, *Rev. Sci. Instr.*, **26**, 765 (1955).

[2] K. S. Lion, *Helv. Physica Acta*, **14**, 21 (1941), and *Rev. Sci. Instr.*, **13**, 338 (1942); a similar system is described by J. F. Steinhaus, *Rev. Sci. Instr.*, **27** 575 (1956).

The spherical shape of the tube permits the measurement of the field strength $|F|$, regardless of the direction; if the direction of the field is of interest, the spherical tube can be replaced by a longitudinal tube. The error in fields that are practically homogeneous over the volume of the sphere (diameter about 1 to 2 cm) is about 2 per cent. The system cannot be used at field strengths below about 20 to 100 volts/cm (depending upon the frequency). The indication is instantaneous.

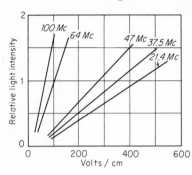

FIG. (4-5)4. Calibration curves of the system shown in Fig. (4-5)3, for different frequencies of the rf field.

### 4-54. THERMAL METHOD IN RF FIELDS

If a high-frequency field is maintained in a conductive medium (electrolyte), power will be absorbed in the medium. The power concentration in watts per cubic centimeter in every volume element of the conductive medium is

$$p = \frac{\sigma \epsilon F^2}{8\pi}$$

where $\sigma$ is the conductivity of the medium, $\epsilon$ its dielectric constant, and $F$ the field strength. The power is converted into heat. If heat exchange between different parts of the medium is negligible, the field-strength distribution in the medium can be determined from local temperature measurements.[1]

[1] J. Paetzold and P. Betz, *Z. ges. exp. Med.*, **94**, 696 (1934); see also K. S. Lion, *Arch. Physical Med.*, **28**, 345 (1947).

# 5

# *Radiation  Transducers*

The field is divided into two sections:  transducers responding to optical radiation (5-1) and transducers responding to ionizing radiation (5-2).

## 5-1. Optical Radiation Transducers

Optical transducers comprise photoemissive systems (5-11), photoconductive (5-12), photovoltaic (5-13), and thermal systems (5-14).

### 5-11. PHOTOEMISSIVE TRANSDUCERS

A photoemissive transducer, Fig. (5-1)1, contains in general a metallic cathode $C$ and an anode $A$ in an evacuated envelope.  Light strikes the cathode and liberates electrons.  The electrons are attracted by and move toward a positive anode and thereby form an electric current.

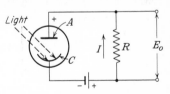

If a light quantum of the wavelength $\lambda$ interacts with an electron bound in a metal, the entire quantum energy $A = hc/\lambda$ ($h$, Planck's constant, $6.62 \times 10^{-34}$ watt/sec$^2$; $c$, propagation velocity of

FIG. (5-1)1. Photoelectric cell and basic circuit.

light, $3 \times 10^8$ m/sec) will be converted into the kinetic energy of the electron.

The kinetic energy enables the electron to leave the metal surface with an energy

$$A' = \frac{hc}{\lambda} - e\varphi$$

233

where $e\varphi$ is the energy required for the electron to overcome the binding forces of the metal, $e$ is the charge of the electron, and $\varphi$ is the work function of the material of the cathode. Of all elements, the smallest values for $\varphi$ are found for the alkali metals, in particular

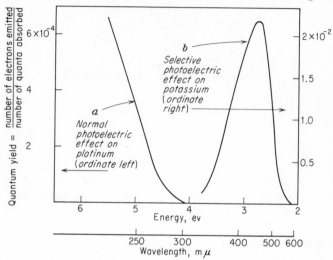

Fig. (5-1)2. Quantum yield for photoelectric emission: (*a*) normal photoelectric effect on a platinum surface; (*b*) selective photoelectric effect on a potassium–oxygen surface (*after R. W. Pohl, Optik, Springer Verlag, Berlin,* 1940; *by permission*).

for cesium; compounds of the alkali metals can have even smaller work functions.

The energy with which the photoelectrically liberated electron leaves the metal becomes zero for

$$\frac{hc}{\lambda_0} = e\varphi$$

$$\lambda_0 = \frac{hc}{e\varphi} = \frac{1{,}240}{\varphi}$$

where $\lambda_0$ is the threshold wavelength (in millimicrons) for a metal surface with the work function $\varphi$. Any radiation with a wavelength greater than $\lambda_0$ will not cause photoemission from this surface. The threshold wavelength of commercial photoelectric surfaces is, in general, between 600 and 1,200 m$\mu$.

The quantum yield, i.e., the number of electrons emitted for a given number of absorbed light quanta, increases with decreasing wavelength from zero at the threshold wavelength, in general, in a monotonic function. A normal photoelectric effect is demonstrated by Fig. (5-1)2, curve $a$. Some metals, e.g., alkali metals in an atmosphere containing oxygen or hydrogen, show a considerably increased yield in a narrow wavelength range. A selective photoelectric effect is shown by Fig. (5-1)2, curve $b$. The selective photoeffect has its origin in a thin surface layer.

The quantum yield on pure metal surfaces is of the order of 0.1 per cent, i.e., one electron emitted per thousand absorbed quanta. Commercial photoelectric devices have a quantum yield of the order of 5 per cent. Quantum yields as high as 30 per cent have been observed and used in photoemissive devices.

The photoelectric effect is not limited to metals but occurs also on insulators and semiconductors.

The photoelectric current $I$ is proportional to the illumination $\Phi$ of the cathode

$$I = S\Phi$$

where $S$ is the sensitivity, a magnitude depending upon the composition of the cathode surface and, to a lesser extent, upon the tube construction. $\Phi$ is the incident radiation flux, expressed either in watts or in lumens. Correspondingly, the sensitivity is expressed either in amperes per watt ("radiant sensitivity") or in amperes per lumen ("luminous sensitivity").

The sensitivity of photoemissive cathodes varies strongly with the wavelength of the incident radiation. The response characteristic $I = f(\lambda)$ depends upon the composition and the processing of the cathode. A considerable amount of work has been done to produce photocathodes of high sensitivity over the entire spectrum or in certain ranges of the spectrum.

For references, see V. K. Zworykin and E. G. Ramberg, "Photoelectricity and Its Application," John Wiley & Sons, Inc., New York, 1949; P. Görlich, "Die lichtelektrischen Zellen," Akademische Verlagsgesellschaft Geest and Portig K.-G., Leipzig, 1951; A. Sommer, "Photoelectric Tubes," Methuen & Co., Ltd., London, and John Wiley & Sons, Inc., New York, 1951; for review papers, see M. Ploke, *Arch. tech. Messen*, J 391–5, November, 1953, and J 391–6, January, 1954.

Three types of photoemissive cathodes are being used:

1. Ag-O-M type (M stands for an alkali metal). The cathode consists essentially of a silver layer that is oxidized and covered with a layer of an alkali metal (e.g., cesium). The cathode has one maximum sensitivity in the region around 800 m$\mu$ and a second maximum in the

ultraviolet region. Examples of response characteristics are shown in Fig. (5-1)3, *S*-1 and *S*-3.

2. Sb-M type ("alloy type"). The maximum response for these cathodes is between 400 and 500 m$\mu$, the threshold wavelength between 600 and 700 m$\mu$. The thermionic dark current (see below) of these cathodes is small. At the response maximum, the quantum

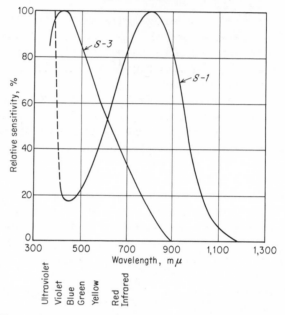

Fig. (5-1)3. Relative sensitivity of different commercial photocathodes (*from RCA Tube Manual; by permission*).

yield is very high, about 20 per cent. Typical response characteristics are shown in Fig. (5-1)4, *S*-4, *S*-5, and *S*-9. Considerable increase of the sensitivity and of the threshold wavelength can be obtained by combining several alkali metals with the antimony.[1]

3. Bi-O-Ag-M type. The maximum of the spectral response curve is in the same wavelength region as that of the Ag-O-M type. A response characteristic is shown in Fig. (5-1)5, *S*-8.

The response characteristic of all photoelectric devices is influenced in the ultraviolet region by absorption in the glass wall of the bulb. Lime glass absorbs practically all radiation of a wavelength less than 300 m$\mu$. Ultraviolet transparent glass (Corex D) transmits down to 280 m$\mu$, and a specially thin window will transmit

[1] A. H. Sommer, *Rev. Sci. Instr.*, **26**, 725 (1955).

70 per cent of the radiation at 250 m$\mu$. Quartz transmits ultraviolet light down to 200 m$\mu$ and begins to absorb substantially at shorter wavelengths. Ultraviolet radiation and radiation of smaller wavelengths can also be measured by converting it into visible light by means of a fluorescent screen and measuring the screen brightness with photoelectric devices that are sensitive to visible radiation.[1]

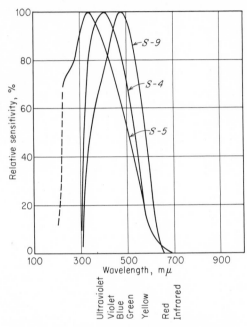

Fig. (5-1)4. Relative sensitivity of different commercial photocathodes (*from RCA Tube Manual; by permission*).

Since the photocathode response varies with the wavelength, the integral magnitude "sensitivity," in amperes per lumen, has only a limited significance. The sensitivity of photoelectric devices can be compared either on the basis of a light source of equal radiant flux at all wavelengths or of a standardized light source with known and reproducible spectral distribution. For the measurement of the luminous sensitivity, it has become standard practice to use a tungsten lamp operated at a filament temperature of 2870°K. Alternatively, 1 hololumen, i.e., the radiant flux of all wavelengths of a tungsten lamp at a color temperature of 2848°K, is used which, evaluated as

[1] R. D. Cowen, *Bull. Am. Phys. Soc.*, **22**, F-7 (December, 1947).

visible light, equals 1 lumen. Sometimes (but rarely) the International Candle is used as a standard source. The International Candle furnishes a radiation of $1.05 \times 10^{-4}$ watt/cm² at a distance of 1 m.

For a maximum output, the peak of the photocathode response characteristic should be matched to that of the light source. The

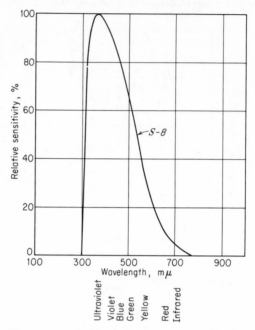

FIG. (5-1)5. Relative sensitivity of different commercial photocathodes (*from RCA Tube Manual; by permission*).

photocathode is not always equally sensitive over its entire surface but may have patches of higher or lower sensitivity. Uniform illumination of the cathode is recommended, therefore.

Undesirable output signals in photoemissive tubes are caused by dark current, i.e., a current passing through the tube even in the absence of light, as well as by shot effect and by Johnson effect. Sources of undesirable signals of minor importance can be flicker effect and wall charges.

The dark current can have four causes: (1) thermionic emission from the cathode and, in photomultipliers, from the dynodes, (2) leakage current over the glass wall, (3) current caused by the impact of positive ions upon the cathode (or, in photomultipliers, the dynodes), and (4) field emission.

The thermionic emission from the cathode can be computed from the Richardson law. If the applied plate voltage is high enough to cause all emitted electrons to move to the anode (no space charge), the thermionically emitted current is

$$i_T = Aat^2 e^{-(e\varphi/Kt)}$$

where $A$ is a constant (for pure metals about $1.2 \times 10^6$ amp/$m^2$ deg $K^2$), $a$ the area of the cathode, $t$ the temperature in degrees Kelvin, $\varphi$ the work function of the emitting metal in volts, $e$ the electron charge in coulombs, and $K$ is Boltzmann's constant. The thermionic emission increases rapidly as $\varphi$ diminishes, i.e., as the threshold wavelength is shifted toward the infrared part of the spectrum. The thermionic emission can be reduced considerably by cooling of the emitting electrode. A refrigerating chamber for this purpose is described by Engstrom.[1]

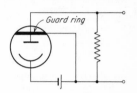

FIG. (5-1)6. Photoelectric cell with guard ring to reduce the effect of leakage current.

Leakage current over the glass wall of the tube can be reduced by reducing the voltage applied between the electrodes or by cleaning and drying the glass surface or coating it with a water-repellent compound. A very good means to avoid the influence of the leakage current in the output consists in the application of a guard ring around one of the electrodes. A circuit diagram for the guard ring arrangement is shown in Fig. (5-1)6.

The current caused by the impact of positive ions upon the cathode and current caused by field emission are, in general, less important than leakage current and thermionic emission. Both can be reduced by reducing the applied voltage; apparently, the current caused by ion impact can be reduced by cooling. Static charges on the glass wall are usually avoided by coating the walls of the tube with a conductive layer.

Several methods have been used to separate the dark current from the photoelectric current: The light source can be modulated or periodically interrupted (light chopper) before entering the photoelectric cell. The resulting photoelectric a-c component in the output from the photoelectric device can be separated from the d-c component, i.e., thermionic and leakage current.

Kalmus and Striker[2] have placed the photoelectric cell in the field

[1] R. W. Engstrom, *Rev. Sci. Instr.*, **18,** 587 (1947).
[2] H. P. Kalmus and G. O. Striker, *Rev. Sci. Instr.*, **19,** 79 (1948).

of an a-c magnet driven with 200 cps, as shown in (5-1)7. The magnetic field deflects the electrons (both photoelectric and thermionic emission) and modulates the electronic current with 400 cycles, while the leakage current remains unmodulated. The a-c component is amplified in a subsequent a-c amplifier. The method permits with simple means the measurement of light intensities down to $3 \times 10^{-7}$ lumen at a signal-to-noise ratio of 1.

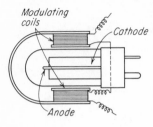

FIG. (5-1)7. Arrangement for the separation of leakage and emission current by magnetic modulation of the emission current [*from H. P. Kalmus and G. O. Striker, Rev. Sci. Instr.*, **19**, 79 (1948); *by permission*].

Shot effect, i.e., the random fluctuation of the electron current through the tube, causes a variational (a-c) current, superimposed on the tube current. The frequency of the variational current is uniformly distributed over the frequency spectrum, but the measurable part of this current is limited by the band width $\Delta f$ of the associated equipment. The mean square value of the current is

$$\bar{I}_s^2 = 3.2 \times 10^{-19} I_0 \, \Delta f$$

where $I_0$ is the average emission current in the tube. If the load resistance connected to the tube is $R$, the mean square value of the noise output voltage caused by shot effect is

$$\bar{E}_s^2 = 3.2 \times 10^{-19} I R^2 \, \Delta f$$

Johnson noise, i.e., a randomly varying potential difference arising between the terminals of a resistor, caused by thermal agitation of the electrons, is

$$\bar{E}_{\mathrm{th}}^2 = 4kt R \, \Delta f$$

At room temperature

$$\bar{E}_{\mathrm{th}}^2 = 1.6 \times 10^{-20} R \, \Delta f$$

where $k$ is the Boltzmann constant, $t$ the absolute temperature of the tube, and $R$ is either the value of the load resistance or of the resistive component of the complex load impedance.

The total noise voltage, comprising shot noise and Johnson noise, is, at room temperature ($t = 293°\mathrm{K}$),

$$E_{N,\mathrm{rms}} = [3.2 \times 10^{-19} R(I R - 0.05) \, \Delta f]^{\frac{1}{2}}$$

*a. Vacuum Photoelectric Cells.* The standard circuit for vacuum photoelectric cells is shown in Fig. (5-1)1. In general, these cells have a moderate sensitivity; the luminous sensitivity is between 10

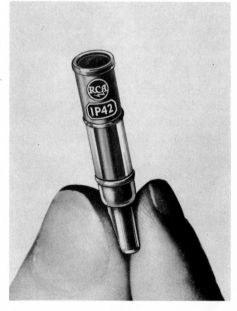

Fig. (5-1)8. Photoelectric cells: (a) low-leakage type with anode terminal cap; (b) twin type; (c) cartridge type; (d) small type for end-on observation (*Radio Corporation of America*).

and 100 $\mu$A/lumen, the radiant sensitivity is between 0.002 and 0.1 $\mu$A/$\mu$W for illumination at peak spectral sensitivity. Vacuum cells are stable and do not change their characteristic over long periods of time, particularly when operated at low plate voltages and protected from excessive light.

Constructions of common photoelectric cells are shown in Figs. (5-1)8a to d. Low values of leakage current (order of magnitude $10^{-10}$

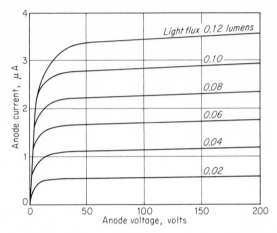

Fig. (5-1)9. Voltage–current characteristic of a vacuum photoelectric cell (*from RCA Tube Manual, Tube 6570; by permission*).

amp) are obtained with types in which the connection to one electrode is made on top of the bulb. The dark current of tubes in which both connections are made through the tube socket is of the order of $10^{-8}$ to $5 \times 10^{-9}$ amp at room temperature and a plate voltage of about 200 volts.

Typical voltage-current characteristics are shown in Fig. (5-1)9. After an initial rise of the plate current with increased plate voltage (transition from space-charge-limited to emission-limited operation) the current remains almost constant with further increase of the plate voltage (saturation region). The small remaining increase of current in the saturation region is brought about by electrons which, because of their initial velocity, bypass the anode, circle around it, and return to the cathode region. The effect is reduced at higher plate voltages.

The current in the saturation region is essentially proportional to the incident light flux. A small deviation from linearity (usually of the order of 1 per cent) is due to field deformation in the tube.

The response time of the vacuum photoelectric cell is limited only by the transit time of the electrons and is of the order of $10^{-9}$ sec. However, the response time of a circuit connected to the photoelectric cell may be increased by the capacitance of the tube (of the order of 1 to 5 $\mu\mu$F) and the resistance of the circuit.

The range of light intensities for which a photoelectric cell can be used is limited at the lower end by noise consideration and dark current (see the general considerations mentioned above) and at the upper end by fatigue effects and stability considerations. The illumination should be such that the emission current for continuous operation does not exceed the order of 1 $\mu$A/cm² of the cathode surface. The output impedance is of the order of several megohms.

The maximum ambient temperatures which can be tolerated are 100°C for tubes with $S$-1, $S$-3, and $S$-8 surfaces and 75°C for $S$-4 and $S$-5 surfaces. For constant sensitivity over long periods, operating voltages of 20 volts or less are desirable, since at high voltages positive-ion bombardment from the residual gases may alter the cathode.

*b. Gas-filled Phototubes.* The sensitivity of a phototube can be increased by amplifying the number of electrons in a gas discharge. An inert gas at a pressure of about 1 mm Hg is introduced and a potential of about 100 volts is applied to the electrodes. The emitted photoelectrons are accelerated in the direction of the field, taking energy from the field and losing some of it by elastic collisions with the gas molecules. Eventually the electron, after attaining an energy of about 20 eV, will ionize an atom and produce a positive ion and a new electron. The two electrons are then accelerated again to produce more electrons; the positive ions are directed to the cathode, where a fraction of them will release further electrons. The resultant amplification of the photocurrent may provide a gain by a factor of 5 to 10. The gain depends upon the applied voltage and, to some extent, upon the tube geometry. The current enhancement by gas filling may be greater, by 20 to 30 per cent, for infrared light (900 m$\mu$) than for red light (660 m$\mu$)[1].

The luminous sensitivity of gas-filled phototubes is between 40 and 150 $\mu$A/lumen, the radiant sensitivity between 0.01 and 0.15 $\mu$A/$\mu$W. The permissible current density is between 10 and 30 $\mu$A/cm² of the cathode surface for continuous operation and up to 600 $\mu$A/cm² for short-time applications.

Voltage-current characteristics of gas-filled phototubes are shown in Fig. (5-1)10. Below the ionization potential of the gas ($\sim$15 volts)

[1] W. S. Huxford, *Phys. Rev.*, **55**, 754 (1939).

the curves are identical with those of vacuum tubes. With further increase of the applied voltage, the current rises with increasing slope. If the voltage is raised too high, a self-sustained glow discharge will set in which may damage the cathode surface. Operation of the tubes at voltages below this critical value, and the insertion of

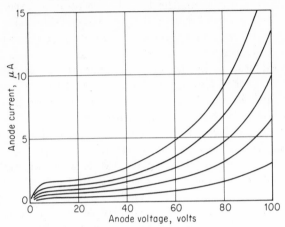

Fig. (5-1)10. Voltage–current characteristic of a gas-filled photoelectric cell (*from RCA Tube Manual, Tube 1P–41; by permission*).

a resistor to limit the current in case of an accidental overvoltage are recommended.

The output current is not strictly a linear function of the light flux; at higher levels of illumination in particular, the curve presenting output current versus incident light flux rises with increased slope.

Since the current in a gas tube is carried by electrons as well as by positive ions, it is reasonable to expect a time lag in the response, which is not observed in a vacuum tube. Figure (5-1)11 shows the current output for a silver-cesium oxide-cesium phototube containing argon at a pressure of 0.15 mm Hg, operated at different voltages and irradiated with light pulses. For light signals modulated sinusoidally at different frequencies, the output drops off at higher frequencies, Fig. (5-1)12. For some applications the drop of the frequency characteristic may be compensated for by subsequent networks.

Dark currents in commercial gas tubes are, in general, of the order of $10^{-8}$ to $10^{-7}$ amp. Gas phototubes are available with the same spectral characteristics in the visible and infrared as vacuum phototubes. Their application is primarily in the field of relay operation and sound reproduction where their slight nonlinearity and lower output at high frequencies play a minor role.  In physical structure they resemble corresponding vacuum phototubes.

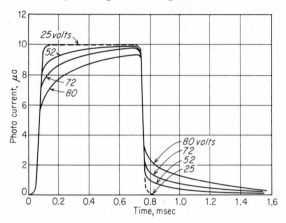

FIG. (5-1)11. Time-response characteristic of a gas-filled photoelectric cell for different plate voltages [*from W. S. Huxford and R. W. Engstrom, Rev. Sci. Instr.,* **8**, 385 (1937);  *by permission*].

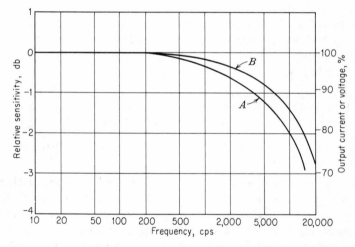

FIG. (5-1)12. Response of a gas-filled photoelectric tube to light modulated at different frequencies: (*A*) phototube having *S*-1 or *S*-3 response, Fig. (5-1)3; (*B*) phototube having *S*-4 response, Fig. (5-1)4 (*from RCA Tube Manual; by permission*).

*c. Photomultipliers.*  A schematic diagram of a photomultiplier is shown in Fig. (5-1)13.  Light strikes a photocathode $C$ and liberates electrons;  the electrons are accelerated by a voltage $E_1$ and focused upon an electrode, the dynode $D_1$, where each incident electron causes the emission of several secondary electrons. The same process is repeated at the dynodes $D_2$, $D_3$, .... . The electrons from the last dynode stage are collected by a positive anode $A$, and the current $I_a$ is measured. If the gain of each stage (the number of electrons

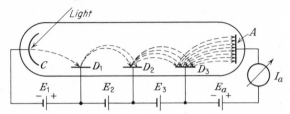

FIG. (5-1)13. Photomultiplier, schematic diagram.

formed by secondary emission for each primary electron) is $g$ and if $n$ dynode stages are used, the total amplification is

$$A = g^n$$

The value of $g$ varies with the voltage between successive dynodes and with the surface composition and the geometry of the dynodes from 0.5 to about 10.

The focusing of the electrons from one stage to the next can be accomplished with magnetic or electrostatic electron-optical systems. Recent constructions use electrostatic focusing systems almost exclusively.[1]

Commercial photomultipliers are built with 9 to 14 stages and furnish a gain between $10^5$ and more than $10^7$ ($10^9$, EMI tube 6262): *Nucleonics,* **10** (3), 34 (1952). Photomultipliers with a greater number of dynodes have been built: 18 stages, with a gain up to $10^{12}$, described by N. Schaetti, *Helv. Physica Acta,* **23**, 108 (1950); 19 stages, by A. Lallemand, *Le Vide,* **4**, 618 (1949). Refer to *Chem. Abstracts,* **44**, 381b (1950).

The gain and the sensitivity of photomultipliers vary with the voltages applied to the dynode stages. A typical characteristic for a single stage showing the gain per stage over a wide range of applied voltages is shown in Fig. (5-1)14; for continuous operation only the part indicated by a solid line is used. In this range, the characteristic can be expressed analytically, according to Larson and Salinger,[2] by

$$g = k\sqrt{E_{st}}$$

[1] For references on the construction of photomultipliers, see G. A. Morton, "The Scintillation Counter," in L. Marton (ed.), "Advances in Electronics," vol. IV, pp. 71ff., Academic Press, Inc., New York, 1952.

[2] C. C. Larson and H. Salinger, *Rev. Sci. Instr.,* **11**, 227 (1940).

where $k$ is a constant and $E_{st}$ is the voltage applied to the stage. A characteristic showing the over-all current amplification of a photo-multiplier as a function of the applied voltage is shown in Fig. (5-1)15.

The amplification can be controlled, within wide limits, by changing the supply voltage applied to the photomultiplier. A further

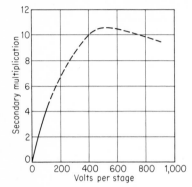

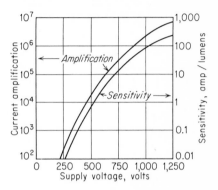

FIG. (5-1)14. Gain per stage of a photo-multiplier (typical for RCA 931 A- and 1P 21-type dynode surfaces)[*from R.F. Post, Nucleonics*, **10**, 46 (May, 1952); *by permission*].

FIG. (5-1)15. Over-all amplification and sensitivity of a photomultiplier, voltage per stage $= \frac{1}{10}$ of supply voltage (*from RCA Tube Manual, Tube* 1P 21; *by permission*).

method to control (reduce) the amplification is to defocus the electron stream in any one of the sections of the tube.

The luminous sensitivity of commercial photomultipliers is generally in the vicinity of 20 amp/lumen but varies widely for different types and even for different specimens of the same type. At its highest, it is 20,000 amps/lumen.[1] The sensitivity characteristic is parallel to that showing the amplification, as illustrated in Fig. (5-1)15. Figures for the radiant sensitivity for normal continuous operation reported in the literature vary between 400 and 80,000 $\mu A/\mu W$ for incident radiation of a wavelength at which the cathode has a maximum sensitivity.

INPUT CHARACTERISTICS. The spectral sensitivities of photomultiplier tubes are determined by the characteristics of their cathodes and are similar to those described in Figs. (5-1)3 to 5. Photomultipliers that are sensitive in the gamma-ray region are described by Graves and Koch.[2] The sensitivity of the photocathodes in photomultipliers is, in general, of the order of 60 $\mu A/lumen$. The light-intensity level for which a photomultiplier is suitable is determined

[1] J. D. McGee, *Nucleonics*, **10**, 34 (March, 1952).

[2] J. D. Graves and G. E. Koch, *Rev. Sci. Instr.*, **21**, 304 (1950).

primarily by the method of operation of the photomultiplier (see below). The lower limit is set by a dark current and by noise. By counting the pulses caused by individual photoelectrons, the lower limit in a 5-min period of observation is $2 \times 10^{-16}$ lumen.[1] At high levels of illumination, the useful range of a photomultiplier is limited by fatigue and lifetime considerations. This limit is, in general, of the order of $10^{-6}$ to $10^{-4}$ lumen for standard photomultipliers and continuous operation but can be raised appreciably for intermittent operation.

An exposure to light in excess of the specified level can cause a reversible and, under extreme conditions, irreversible reduction of the sensitivity (fatigue). The fatigue effect can be caused, even without the application of a voltage to the photomultiplier, by an attachment of electronegative gases that remain in the tube after pumping.[2] With voltage applied to the photomultiplier, the impact of positive ions tends to deteriorate the cathode surface or the dynode surfaces. Also, the exposure to high levels of illumination causes electrons to be emitted from the cathode at a high rate. These electrons are not replenished fast enough from internal layers, and a number of positive ions remain unneutralized.[3] The result is a decrease of sensitivity and may be a shift of spectral distribution, in particular a reduction of the photoelectric threshold to shorter wavelengths. Frequently, the photocathode recovers after a period of idleness (several minutes to hours) in darkness at room temperature; recovery may also be achieved by short heating or by irradiation with red or infrared light of a wavelength longer than the threshold wavelength. A fatigue effect also occurs at the last dynode stage or stages under the influence of high current. If the continuous anode current is less than 1 mA, the tube usually recovers after storing in darkness for a day. At higher currents, the sensitivity can be irreversibly lost.[4] The effect is probably caused by a dissociation of surface molecules at the last dynode and a subsequent diffusion of the cesium off the surface.[5]

Commercial photomultipliers are available with opaque or transparent photocathodes. Tubes with transparent photocathodes are usually built for end-on observation so that the cathode can be brought close to the illuminating source (e.g., scintillation crystal).

[1] R. W. Engstrom, *J. Opt. Soc. Am.*, **37**, 420 (1947).

[2] J. H. de Boer, "Electron Emission and Adsorption Phenomena," chap. 6, The Macmillan Company, New York, and Cambridge University Press, London, 1935.

[3] de Boer, *ibid.*, chap. 8, sec. 96.

[4] Engstrom, *loc. cit.*,

[5] Report on a paper by L. Cathey, *Nucleonics*, **16**, 59 (June, 1958).

For the same application, windows and photocathodes of large dimensions ($>4$ in. in diameter, more than 15 in.$^2$ of area) have been built. The sensitivity of the cathode is not always constant over its entire surface but there may be zones of higher or lower sensitivity.

OUTPUT CHARACTERISTICS. For continuous operation and if long-time stability is required, the output current should be limited to a value of $10^{-5}$ to $10^{-4}$ amp. For a short duration of not more than 30 sec an average current of $10^{-3}$ or even $10^{-2}$ amp is allowable. Much higher currents can be used for pulsed operation. The operation of a tube furnishing output-current pulses as high as 0.2 to 0.3 amp is described by Greenblatt;[1] a tube for pulsed operation with an output current of the order of 1 amp is described by Post;[2] and Singer, Neher, and Ruehle[3] have operated photomultiplier tubes with output-current pulses as high as 15 amp.

The output impedance is essentially that of the load resistor in the anode circuit; it is, in general, in the megohm range, but for pulse operation values as low as 100 ohms have been used.[4]

The output from a photomultiplier is a linear function of the incident light. The deviations from linearity are about 3 per cent for a variation of the light flux between $10^{-9}$ and $10^{-3}$ lumen. Nonlinearity resulting from space charges in the last stages may set in at higher light levels. Such space charges frequently occur not at the anode stage but at the third or second-to-the-last stage. Nonlinearity at low levels (scintillation counters) may be due to statistical variations.[5] Nonlinearity of photomultipliers caused by the action of the voltage divider which supplies the partial voltages for the dynode stages is discussed below under Circuits for Photomultipliers.

The response time of a photomultiplier is, in general, shorter than $10^{-8}$ sec. In the time interval between $10^{-9}$ to $10^{-8}$ sec, dynamic errors begin to become noticeable. Primarily two effects can be observed: a time delay between input and output signal and a spread of the transit time. A short light pulse will cause a longer pulse of anode current (time dispersion). The time delay between input and output signal is due to the finite transit time of the electrons through the tube. The time dispersion is due (1) to differences in the initial velocities of the emitted photoelectrons, (2) to differences in path lengths of the electrons, and (3) to a lesser extent, to

[1] M. H. Greenblatt et al., *Nucleonics*, **10**, 44 (August, 1952).

[2] R. F. Post, *Nucleonics*, **10**, 46 (May, 1952).

[3] S. Singer, L. K. Neher, and R. A. Ruehle, *Rev. Sci. Instr.*, **27**, 40 (1956).

[4] Post, *loc. cit.*

[5] G. A. Morton, *op. cit.*, p. 78.

fluctuating time lags in secondary emission. This latter time dispersion is probably of the order of $10^{-11}$ to $10^{-10}$ sec. Sard,[1] in a theoretical study, has shown that for a single electron a current pulse in the anode circuit of a duration of about $6 \times 10^{-10}$ sec is to be expected. Practically observed time spreads under normal operating

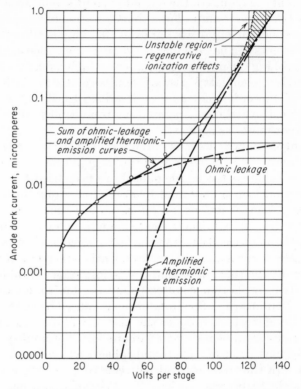

FIG. (5-1)16. Dark current in a photomultiplier tube [*from R. W. Engstrom, Rev. Sci. Instr.*, **37**, 420 (1947); *by permission*].

conditions are of the same order of magnitude. The effect of the different initial velocities of the photoelectrons can be reduced by first accelerating the electrons when they leave the cathode and decelerating them before they reach the first dynode; the effect of varying path lengths can be overcome by the use of a spherical cathode surface and spherical electron-optical geometry behind the cathode. Tubes having these features show time dispersions of less

---

[1] R. D. Sard, *J. Appl. Physics*, **17**, 768 (1947).

than $10^{-10}$ sec.[1]  The time dispersion decreases approximately with the square root of the applied voltage per stage and can be reduced, therefore, by operating the photomultiplier at higher than normal voltage (see Pulsed Operation, below).

Undesirable signals can be caused by dark current and noise. Dark current, by itself, does not necessarily introduce an error if corrections for it can be applied.  However, the noise component of the dark current is frequently objection-able. The dark current in commercial photomultipliers at room temperature is generally of the order of $10^{-7}$ amp.

The dark current is caused by three phenomena: (1) leakage current, predominantly at the inside of the tube and frequently caused by traces of cesium, but also occurring at the outside of the tube and in the tube socket;  (2) thermionic emission from the photocathode and, to a smaller extent, from the first dynode stages, and (3) regenerative ionization, i.e., the impact of positive ions upon the

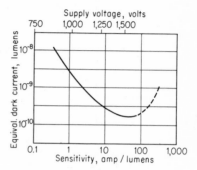

FIG. (5-1)17. Light input equivalent to the dark current of a photomultiplier (*from RCA Tube Manual, Tube 6342; by permission*).

cathode, but also the incidence of light from excited gas molecules and from glass fluorescence upon the cathode.  The relative magnitude of these three sources of dark current and their variation with the voltage applied per stage are shown in Fig. (5-1)16.

The dark current is frequently expressed by the equivalent light input, i.e., by that amount of light at the cathode that causes an anode current of the same magnitude as the dark current. The equivalent light level is proportional to the dark current but inversely proportional to the sensitivity.  If plotted against the sensitivity, which increases with increasing voltage per stage, a curve like that shown in Fig. (5-1)17 will ensue.

Leakage current on the outside can be reduced in some cases by coating the outside of the tube with a conductive layer and applying appropriate potentials.[2]

The thermionic component of the dark current can be reduced by

[1] Anon., Report on the Sixth Scintillation Counter Symposium, Washington, *Nucleonics*, **16**, 58–59 (June, 1958).

[2] Z. Náray, *Acta Physica Acad. Sci. Hung.*, **5**, 159 (1955), and H. J. Marrinan, *J. Opt. Soc. Am.*, **43**, 1211 (1953).

cooling of the tube. Roughly, each 10°C decrease of temperature causes a reduction of the thermionic current to one-half. The thermionic current may also be reduced by reducing the sensitive area of the photo-cathode, as described by Náray.[1] By electron-optical

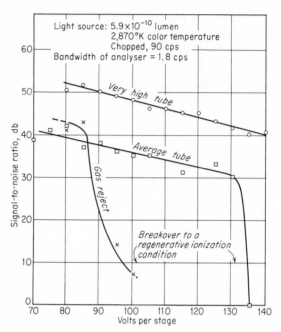

Fig. (5-1)18. Noise output from three photomultiplier tubes as a function of the voltage applied per stage [*from R. W. Engstrom, Rev. Sci. Instr.*, **37**, 420 (1947); *by permission*].

means an image of a small area of the photocathode is formed on the first dynode. Thermal electrons from outside this area do not reach the dynode, and the anode current due to thermionic emission is reduced by a factor of 100.[2]

Noise in photomultipliers is of twofold origin: the random variation of the thermionic emission and that of the photoemission. The rms noise current, measured at the anode, is

$$(\bar{I}^2)^{\frac{1}{2}} = G[2e\,\Delta f(i_t + i_p)]^{\frac{1}{2}}$$

where $G$ is the amplification (gain) of the photomultiplier tube, $e$ the charge of the electron, $\Delta f$ the bandwidth of the detecting system, $i_t$ the thermal emission from the cathode (at room temperature of the

---

[1] Náray, *loc. cit.*

[2] See also G. E. Kron, *Astrophys. J.*, **103**, 324 (1946).

order of $10^{-14}$ amp), and $i_p$ the photocurrent at the cathode. The signal-to-noise ratio is

$$\frac{S}{N} = \left(\frac{i_p^2}{2e\,\Delta f(i_t + i_p)}\right)^{\frac{1}{2}} \tag{1}$$

Engstrom[1] remarks that this equation does not include the noise resulting from thermionic emission from the dynodes, which he estimates to be 3 per cent of that expressed in Eq. (1), nor that associated with random secondary emission, of the order of 15 per cent.

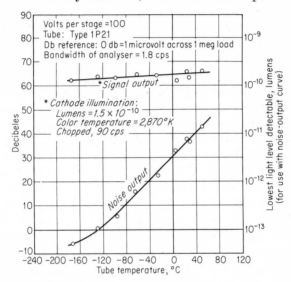

FIG. (5-1)19. Variation of signal and noise with the temperature of the photomultiplier. The signal-to-noise ratio can be found as the difference between both curves [*from R. W. Engstrom, Rev. Sci. Instr.,* **37**, 420 (1947); *by permission*].

From Eq. (1) one would expect that the signal-to-noise ratio is independent of the amplification and, therefore, independent of the voltage applied per stage. However, this is not so in practice. Experimentally obtained characteristics, shown in Fig. (5-1)18, indicate that the signal-to-noise ratio decreases slightly with increasing voltage per stage until, at a voltage where regenerative ionization sets in, the $S/N$ ratio declines sharply.

The most effective means to reduce the noise level in photomultipliers consists in reducing the thermionic emission by operating the tube at low temperature. The effect of temperature on the signal output and the noise output is illustrated in Fig. (5-1)19. Reduction

[1] Engstrom, *loc. cit.*

of the temperature from room temperature to that of liquid air reduces the noise level by a factor of 100.

The equivalent noise input, i.e., that light intensity at the input of the photomultiplier which furnishes at its output a current equal to the rms noise output over a band width of 1 cycle is, for commercial photomultipliers operated at room temperature, between $5 \times 10^{-13}$ and $10^{-11}$ lumen.

The noise spectrum is flat over a frequency range from 0 to $10^8$ cps, but falls off between $10^8$ and $10^9$ cps.[1]  Gordon and Hodgson[2] have observed a reduction of the noise level by a factor of 7 in a photomultiplier that was kept in darkness for five weeks.  The effect seems to be due to a decay of glass fluorescence.

Comparison of a photomultiplier with a photoelectric cell plus amplifier furnishes the following result:  Both the photoelectric cell and the photomultiplier exhibit thermionic emission noise and shot noise, which appear amplified at the output.  However, the amplifier following a photoelectric cell and the coupling resistor between the photoelectric cell and the amplifier also cause a noise component (Johnson and shot noise) in the output.  This noise from the amplifier is constant;  it does not diminish if the light intensity decreases.  Consequently, the signal-to-noise ratio for a photocell plus amplifier decreases for small light intensities.  On the other hand, the shot noise from a photomultiplier is proportional to $\sqrt{i_p}$, therefore proportional to the root of the incident light intensity, so that the shot noise decreases more and more with a decrease of the incident light intensity and only thermionic emission remains as a limiting factor.  However, the superiority of the photomultiplier exists only at small photoelectric currents.  At higher levels of light, when the photocurrent from the cathode reaches the order of $10^{-6}$ amp, the signal-to-noise ratio is almost the same for both systems.[3]

Investigations of the influence of temperature upon the performance of photomultiplier tubes have led to contradictory results.  While some authors find a slight increase of amplification with increasing temperature, as shown, for instance, in Fig. (5-1)19, others have observed a decrease of amplification, sometimes up to 40 per cent for a change of temperature from $-15$ to $-50°C$.[4]

The movement of the electrons can be strongly influenced by

[1] Sard, *loc. cit.*

[2] B. E. Gordon and T. S. Hodgson, *Nucleonics*, **14**, 64 (1956).

[3] J. A. Rajchman and R. L. Snyder, *Electronics*, **13**, 20 (December, 1940).

[4] F. E. Kinard, *Nucleonics*, **15**, 92 (April, 1957);  further references in the same paper.

stray electric and magnetic fields. Even the magnetic field of the earth can cause variations of the photomultiplier output. Adequate shielding should be provided where such stray fields cause errors. Stray light entering the tube can cause erratic results.

OPERATION OF PHOTOMULTIPLIERS. Three different methods of operating photomultipliers are generally distinguished: (1) The light flux falling upon the cathode can be continuous; the output from the photomultiplier is read directly or after d-c amplification. This

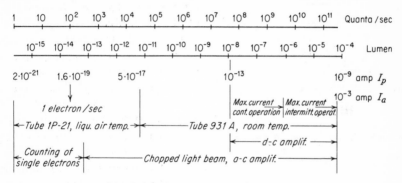

$I_p$ = Photo-(cathode) current      $I_a$ = Anode (output) current

FIG. (5-1)20. Operating ranges for photomultipliers [*from M. Ploke, Arch. tech. Messen, J 391–6, January 1954; by permission*].

method is generally applicable when the incident light level is higher than $10^{-8}$ lumen. At lower levels, the unavoidable drift of d-c amplifiers can cause serious errors. (2) The light beam can be periodically interrupted or modulated by alternating current, and the output can be fed into a-c amplifiers, which are inherently more stable. The method has the further advantage that the band width of the output-signal detector can be restricted so that the noise level is reduced. The method is in general applicable for light levels as low as $10^{-13}$ lumen. (3) For lower light levels the bursts caused by single photoelectrons are observed or counted. This method permits the measurement of smallest light levels, down to the order of $2 \times 10^{-16}$ lumen. A synopsis of the different methods of operation for photomultiplier tubes is shown in Fig. (5-1)20.

CIRCUITS FOR PHOTOMULTIPLIERS. The simplest circuit to obtain the appropriate voltages for the single stages is a resistive voltage divider, as shown in Fig. (5-1)21. The operating voltage applied to the voltage divider is of the order of 1 kV; for d-c output, the positive pole is preferably grounded; the partial voltages for the single stages are between 70 and 150 volts; the output signal $E_o$ appears across the load resistance $R_L$ and is usually amplified in successive stages. Higher voltages are frequently applied to the initial stage to improve the collection of photoelectrons emitted from the cathode; high collection efficiency (95 to 100 per cent) at this stage is important to improve the signal-to-noise ratio. In photomultipliers with a large number of stages ($n > 10$), the last stage or stages also are frequently operated at higher voltages to reduce space charges. An example of a nonuniform voltage distribution and its effect on pulse-height distribution curves is given by Greenblatt.[1]

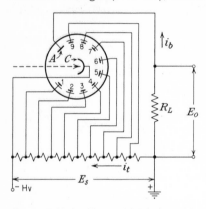

FIG. (5-1)21. Photomultiplier with resistive voltage divider: $C$, cathode; $A$, anode; 1 to 9, dynodes; $E_S$, supply voltage; $E_O$, output voltage across load resistor $R_L$.

The output signal follows a linear function of the incident light flux for small and moderate light levels when the current $i_b$ through the photomultiplier tube is small compared to the current $i_t$ through the resistive network. If the incident light flux increases, the current through the tube increases too, in particular that between the last dynode and the anode. If this current approaches the voltage divider current $i_t$, the voltage distribution of the voltage-dividing network will be upset; the voltages between the last stages will be diminished, those between the initial stages will be increased (under some conditions to the extent that an internal breakdown occurs). The result may be an increase or decrease of the gain as shown in Fig. (5-1)22. The nonlinearity may be useful; for instance, the increase of gain in the region $A$ may be desirable, or the output-current limitation in the region $B$ may protect the photomultiplier tube from damage caused by exposure to excessive light. A circuit with approximately linear response where such protection is enhanced by the use of series resistors is shown in Fig. (5-1)23.[2] The transfer characteristic of a photomultiplier used in connection with this circuit is not entirely linear and has peaks and valleys caused by electrostatic defocusing.

Since the amplification of photomultiplier tubes varies strongly with the supply voltage, operation at constant voltage and the use of

[1] Greenblatt et al., *op. cit.*, p. 47.

[2] R. W. Engstrom and E. Fisher, *Rev. Sci. Instr.*, **28**, 525 (1957).

stabilized supply sources is required.  For instance, a variation of the supply voltage by 1 per cent can cause a variation of the photomultiplier output by as much as 10 per cent.

Numerous methods have been devised to avoid the variation with load of the partial voltages incurred with resistive voltage dividers.

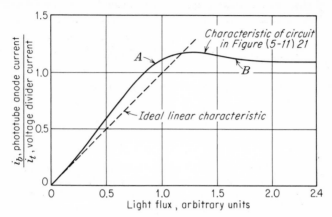

FIG. (5-1)22.  Relative response of an RCA 931-A photomultiplier as a function of the light flux in the circuit of Fig. (5-1)21.  $i_b$, anode current;  $i_{t_0}$, voltage divider current for zero light flux [*from R. W. Engstrom and E. Fischer, Rev. Sci. Instr.*, **28**, 525 (1957); *by permission*].

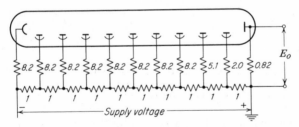

FIG. (5-1)23.  Compensated resistive voltage divider;  all resistance values in megohms [*from R. W. Engstrom and E. Fischer, Rev. Sci. Instr.*, **28**, 525 (1957); *by permission*].

For pulsed light (e.g., scintillation counters), the insertion of capacitors in parallel to the resistors of the dividing network, as shown in Fig. (5-1)24, is frequently satisfactory.  Only the last few stages need, in general, such a capacitive bypass.[1]

[1] F. H. Marshall, J. W. Coltman, and A. I. Bennett, *Rev. Sci. Instr.*, **19**, 744 (1948);  R. Dehn, *Brit. J. Appl. Physics*, **7**, 144 (1956);  J. A. Rajchman and R. L. Snyder, *Electronics*, **13**, 20 (December, 1940).

Voltage-stabilizing elements, such as gas-discharge stabilizers (VR tubes) in voltage-dividing networks, have been described by Mackay and Soule[1] and by Stump and Talley.[2] The use of Zener diodes is described by Flagge and Harris.[3] A circuit diagram of the Zener-diode stabilized voltage divider is shown in Fig. (5-1)25. An improved circuit of this type is described by Hendrick.[4] The methods do not

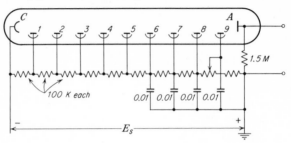

FIG. (5-1)24. Resistive voltage divider with parallel capacitors in the last four stages [*from F. H. Marshall, J. W. Coltman, and A. I. Bennett, Rev. Sci. Instr.,* **19,** 744 (1948); *by permission*].

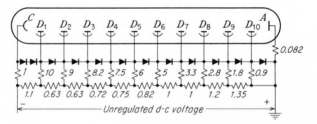

FIG. (5-1)25. Zener diode stabilized voltage divider; all resistance values in megohms [*from B. d'E. Flagge and O. R. Harris, Rev. Sci. Instr.,* **26,** 619 (1955); *by permission*].

permit a variation of the partial voltages to allow for optimum focusing or gain control. This drawback is overcome in a voltage divider described by Kane,[5] who uses triodes as voltage-stabilizing elements. The dynode voltages remain practically constant for an anode current variation from zero to about 6 mA. A resistive voltage divider in conjunction with an auxiliary battery which affords considerable stabilization but reduces the gain to about one-fifth that

[1] A. S. Mackay and R. R. Soule, *Electronics,* **22,** 101 (January, 1949).

[2] R. Stump and H. E. Talley, *Rev. Sci. Instr.,* **25,** 1132 (1954).

[3] B. d'E. Flagge and O. R. Harris, *Rev. Sci. Instr.,* **26,** 619 (1955).

[4] R. W. Hendrick, *Rev. Sci. Instr.,* **27,** 240 (1956).

[5] J. V. Kane, *Rev. Sci. Instr.,* **28,** 582 (1957).

obtained with the standard resistive voltage divider (at the same supply voltage) is described by Sherr and Gerhart.[1]

If continuous operation of the photomultiplier is not essential, the multiplier may be operated at alternating current of line frequency, or higher frequencies, or even at radiofrequencies. Inductive voltage dividers and circuits for a-c operation are described by Zworykin and Ramberg.[2] Since the sensitivity of the photomultiplier increases strongly with applied voltages, the photomultipliers supplied with alternating current will be operative only during the short times when the a-c voltage is near the positive peak value. For this reason, the signal-to-noise ratio of an a-c operated photomultiplier is less than for d-c operation.

PULSED OPERATION. The operation of photomultipliers at high supply voltages is desirable because it improves the amplification, the current output, and the time-dispersion of the amplified pulses. Frequently, the applicable voltage is limited by the effects of ion impact upon the cathode or the earlier dynode stages, leading to noise, instability, and breakdown of the tubes. Since the ion mobility is small compared to that of the electrons, these undesirable effects can be reduced by application of the supply voltage for very short times. Post,[3] has operated photomultipliers with pulses of a duration of $2.5 \times 10^{-6}$ sec or less (repetition rate, 60 per second) and a voltage of four to five times the maximum specified value for d-c operation and has increased the over-all amplification from $5 \times 10^5$ to $10^9$. About 10 to 20 per cent of the commercial tubes investigated could withstand this method of operation. The operation of tubes where the application of high voltage causes field emission cannot be improved by pulsed operation. Singer, Neher, and Ruehle[4] have operated photomultipliers with pulses of $10^{-7}$ sec duration and an over-all dynode voltage of 10 kV.

PHOTOMULTIPLIER CIRCUITS FOR LOGARITHMIC CHARACTERISTIC. The absorption of light follows an exponential law; the optical density $D$ of a light-absorbing layer is defined by

$$D = \log_{10} \frac{I_0}{I}$$

[1] R. Sherr and J. B. Gerhart, *Rev. Sci. Instr.*, **23**, 770 (1952).

[2] V. K. Zworykin and E. G. Ramberg, "Photoelectricity," pp. 259 and 260, John Wiley & Sons, Inc., New York, and Chapman & Hall, Ltd., London, 1956.

[3] Post, *loc. cit.*

[4] Singer, Neher, and Ruehle, *loc. cit.*

where $I_0$ is the intensity of the incident light, and $I$ that of the transmitted light. It is sometimes desirable to obtain an output signal
that is proportional to the logarithm of the incident light flux, for
instance, if the optical density $D$ is to be displayed on a linear scale.

A method to operate a photomultiplier with a logarithmic characteristic has been described by Sweet[1] and is illustrated in Fig. (5-1)26.

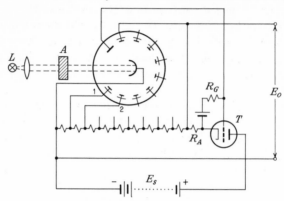

Fig. (5-1)26. Circuit for the operation of a photomultiplier with a logarithmic characteristic [*from
M. H. Sweet, J. Opt. Soc. Am.*, **37**, 432 (1947); *by
permission*].

First, the direct light from the source $L$ (without absorber) is permitted to fall upon the photomultiplier; the illumination intensity
at the cathode is $I_0$ and the observed current at the photomultiplier
output is $i_0$. Second, an absorber $A$ is brought between the source $L$
and the photomultiplier; the light intensity at the cathode is reduced
to the level $I$ and causes a lower output current $i$. The supply voltage
to the photomultiplier is then raised by the amount $\Delta E$ until the
former value of the output current $i_0$ is reached again. The sensitivity
of the tube, i.e., the current for a given illumination, is nearly an
exponential function of the supply voltage, as shown in Fig. (5-1)15;
therefore, the variation of the supply voltage $\Delta E$ at constant output
current is nearly inversely proportional to the logarithm of the illumination intensity, and a meter measuring the applied voltage can
be calibrated directly in terms of the optical density $D$ of the absorber. Measurements of the optical density usually extend over a
range of $D$ from 0 to 3, corresponding to a range of illumination
intensities from 1 to 1,000. Over this range the characteristic deviates from a straight line (correct logarithmic function) by about 10

[1] M. H. Sweet, *J. Opt. Soc. Am.*, **37**, 432 (1947).

to 20 per cent. This deviation can be corrected, for instance, by appropriate division of the scale of the indicating meter.

The circuit shown in Fig. (5-1)26 automatically keeps the output current nearly constant. An increase of the photomultiplier anode current causes the grid of the triode to assume a more negative potential, resulting in a reduction of the voltage applied to the resistive voltage divider and the different stages of the photomultiplier.

Logarithmic densitometers composed from photoelectric cells and elements with logarithmic characteristics are also described by Miller,[1] and by Tiedman.[2]

For review articles and references on photomultipliers, see S. Rodda, "Photoelectric Multipliers," McDonald & Co., Ltd., London, 1953; Zworykin and Ramberg, *op. cit.*, chap. 8; Morton, *op. cit.*; Görlich, *loc. cit.*

*d. Photocounters.* The Geiger counter (5-21*b*) is essentially a photoelectric receiver, which is normally limited to very energetic quanta which cause photoionization and the liberation of electrons either in the walls of the counter or in the gas volume. A counter for ultraviolet and visible light has first been described by Rajewsky.[3]

A schematic diagram of the light counter is shown in Fig. (5-1)27. A cylindrical tube *C* has a window *W* that permits ultraviolet and

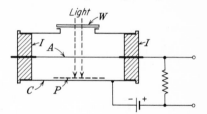

Fig. (5-1)27. Photocounter, schematic diagram.

visible light to enter the inside of the counter. The anode *A* consists of a fine wire held in the center of the tube by means of two insulators *I*. The counter is filled with a suitable gas plus a quenching agent of a pressure of a few centimeters Hg. A photoemissive layer *P* is arranged opposite the window. Electrons liberated from the photoemissive layer migrate toward the anode and cause ionization by collision on their way to the anode. In its original form the counter could be used only for the detection of ultraviolet incident light. Rajewsky indicates that he was capable of measuring as little as 12 quanta/(cm$^2$)(sec) in an observation time of 10 to 12 min. The quantum yield is low, about $10^{-3}$ to $10^{-4}$ electron/quantum.

Different authors have tried to improve the quantum yield and

[1] C. W. Miller, *Rev. Sci. Instr.*, **6**, 125 (1953).

[2] J. A. Tiedman, *Electronics*, **14**, 48 (March, 1941).

[3] B. Rajewsky, *Z. Physik*, **63**, 576 (1930); *Physikal. Z.*, **32**, 121 (1931); *Ann. Physik*, **20** (5), 13 (1934).

the spectral response of the photocounter. Locher[1] has investigated a great number of photoelectric surfaces, liquids, solids, metals, and insulators and has apparently succeeded in covering a spectral range from 90 to 750 m$\mu$.

Further work in this direction is described by K. H. Kreuchen, *Z. Physik,* **94,** 549 (1935), and **97,** 625 (1936); W. Christoph, *Ann. Physik,* **23,** 747 (1935), and *Physikal. Z.,* **37,** 265 (1936); O. S. Duffendack and W. E. Morris, *J. Opt. Soc. Am.,* **32,** 8 (1942); M. V. Scherb, *Phys. Rev.,* **73,** 86 (1948).

The greatest difficulty is the interaction of the gas in the counter with the photoelectric surface.[2] Bauer[3] and Kiepenheuer[4] have tried to overcome this difficulty by separating the counter from the photoelectric system by means of thin metal foils or glass membranes. The point counter has also been used for the measurement of small light intensities.[5]

For review papers, see H. Neuert, *Arch. tech. Messen,* V 422, March, 1942, and C. E. Mandeville and M. V. Scherb, *Nucleonics,* **7,** 34 (November, 1950).

### 5-12. Photoconductive Transducers

Almost all semiconductors are photoconductive, i.e., their electrical resistance decreases when they are exposed to light. The effect

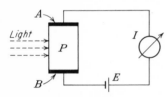

FIG. (5-1)28. Photoconductor, basic circuit.

is particularly strong in selenium, in metal sulfides, oxides, and halides, as well as in germanium and silicon. The basic transducer circuit is shown in Fig. (5-1)28. While the photoconductive layer $P$ is kept in darkness, its resistance is high, and only a small dark current will be observed. Exposure to light causes a decrease of resistance and, consequently, an increase of the current (photocurrent).

A review of the physical processes in photoconduction is given by Rose.[6] The photoconductive process is initiated by the absorption of an incident photon which causes the excitation of an electron or other carrier to the level of the conduction band. The carrier is then

[1] G. L. Locher, *Phys. Rev.,* **53,** 333 (1938), and *Phys. Rev.,* **42,** 525 (1932).

[2] G. W. Barnes and L. H. du Bridge, *Phys. Rev.,* **49,** 409 (1936).

[3] H. Bauer, *Z. Physik,* **71,** 532 (1931).

[4] K. O. Kiepenheuer, *Z. Physik,* **107,** 145 (1937).

[5] H. Bauer and B. Sturm, *Z. Physik,* **94,** 85 (1935).

[6] A. Rose, in R. G. Breckenridge, B. R. Russell, and E. E. Hahn (eds.), "Photoconductivity Conference," John Wiley & Sons, Inc., New York, and Chapman & Hall, Ltd., London, 1956.

free to move through the crystal lattice and form a current. It remains in this state for the duration $\tau$, the statistical average lifetime of the carrier in the conduction band. The total increase of the number of free carriers, caused by $F$ excitations/sec, is

$$N = \tau F$$

The lifetime $\tau$ of the carriers is determined by their rate of recombination; the lifetime of commonly used photoconductors ranges from $10^{-5}$ to $10^{-2}$ sec.

The photocurrent in a photoconductor can be expressed by

$$I_p = \frac{eF\tau}{T_r}$$

where $T_r$ is the transit time of the carriers between the electrodes. If an ohmic contact exists between the electrodes and the photoconductor (i.e., in the absence of a barrier layer), new carriers can move into the photoconductor as the primary carriers move through the photoconductor and out of the space between the electrodes. Since these new carriers take the place of the former, the carriers can be considered to be still alive. The gain of the photoconductor, i.e., the number of charges passing through the photoconductor per photon excitation, is

$$g = \frac{\tau}{T_r}$$

The gain can range from less than unity to several orders of magnitude higher than unity. Gains of more than $10^4$ have been observed.[1]

TRANSFER FUNCTION. In general, the resistivity decreases and the photocurrent increases with increasing incident light flux. The transfer function, i.e., the characteristic of photocurrent versus incident light intensity, can follow a wide variety of patterns; it can increase linearly with the light intensity $\Phi$, or it can follow a $\Phi^n$ law in which the exponent $n$ can be smaller than unity (sublinearity, $n$ between 0.5 and 1). It can also be larger than unity (superlinearity). The situation is further complicated by the fact that the magnitude of $n$ not only depends upon the photoconductor but also upon the level of incident light intensity and upon the temperature. Nonlinearity of the transfer function can appear if the photoconductive transducer is connected in series with a load resistor, so that the partial voltage applied to the photoconductor changes with the level of illumination. A feedback circuit which improves the linearity of the transfer function in such cases is described by Rittner.[2]

[1] R. W. Smith, *Phys. Rev.*, **97**, 1525 (1955).
[2] E. S. Rittner, *Rev. Sci. Instr.*, **18**, 36 (1947).

Photoconductors can respond to radiation over a spectral range extending from thermal radiation through the infrared, visible, ultraviolet spectrum into the range of X rays and gamma rays; some of the photoconductors also respond to bombardment with $\alpha$ rays and electrons. The spectral response of a photoconductive transducer drops sharply at longer wavelengths. The "cutoff wavelength" (i.e., either the wavelength at which the output is equal to the noise level or, in the English literature, that wavelength at which the output is decreased to 50 per cent of its maximum value) depends upon the temperature. Cooling of the photoconductor generally increases its response to longer waves.

The radiation sensitivity of commercial photoconductive transducers is of the order of 300 $\mu A/\mu W$ for a radiation of a wavelength at which the photoconductor has a maximum sensitivity. The luminous sensitivity of photoconductive transducers varies widely, from about $10^{-4}$ to 10 amp/lumen. The sensitivity depends, of course, upon the applied voltage (and upon many other parameters), so that comparisons of photoconductive devices on the basis of sensitivity only are of little value. A standardized method to evaluate a photoconductive cell or to compare the performance of different cells has been proposed by Jones.[1]

In general, the photoconductor current increases linearly with the applied voltage. Polarization phenomena or barrier layers may cause an apparent deviation from linearity.[2] An exponential increase of current with the applied voltage can also be due to space-charge limitation.[3] Nonlinearity of the exponential-increase characteristic is particularly pronounced in photoconductive powders. Thomsen and Bube[4] have found, in cadmium sulfide and cadmium selenide layers produced from microcrystalline powder in a plastic solution, that at low field strength the current varies with the fourth or fifth power of the field, but that at fields greater than 4,000 volts/cm the current-field relationship is linear. The apparent reason for this effect is that at low field strengths the current is limited by the high resistance of the plastic layers between the much less resistant photoconductor granules. This barrier effect is reduced as the voltage increases and the voltage-current characteristic approaches a linear relationship.

The photocurrent does not follow instantaneously a variation of

[1] R. C. Jones, *Rev. Sci. Instr.*, **24**, 1035 (1953).

[2] R. W. Smith, and A. Rose, *Phys. Rev.*, **92**, 857 (1953).

[3] A. Rose and R. W. Smith, *Phys. Rev.*, **92**, 857 (1953).

[4] S. M. Thomsen and R. H. Bube, *Rev. Sci. Instr.*, **26**, 664 (1955).

the illumination, but rises gradually after the excitation is applied and decays gradually after the excitation is removed. The time constants can vary for different photoconductors from microseconds to several minutes; also both time constants decrease with the level of illumination and increase with the temperature. A typical characteristic for a commercial cadmium sulfide cell is shown in Fig. (5-1)29. Rise and decay times of photoconductor currents are not identical

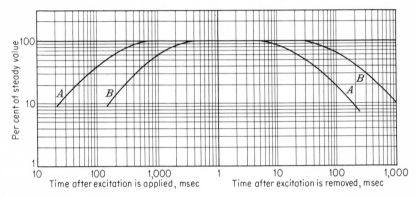

FIG. (5-1)29. Rise and decay of a photoconductive cell (*from RCA Tube Manual, Type* 6957; *by permission*).

with the lifetime $\tau$ of the free carriers but are usually much longer. They are determined by the time required for filling or emptying the lower levels of energy.

Noise in photoconductors is intrinsically very small and consists of photocurrent noise and thermal-agitation noise. In addition to these two components, noise can result from thermally generated dark currents. The contribution of this source to the total noise level is complex; it is possible that the thermally generated "dark" electrons contribute comparatively little to the noise level. It is generally observed, however, that noise in photoconductors can be larger than that caused by the sources mentioned above and that the mean square noise current per unit band width is not equally distributed over the entire frequency spectrum but is proportional to $1/f$, i.e., decreases with increased frequency. It is likely that the source of this excess noise is in the barrier surfaces formed primarily at the contact electrodes. Shulman, Smith, and Rose[1] have succeeded in reducing the noise from cadmium sulfide crystals by orders of magnitude by replacing the silver contact electrodes with indium or gallium

[1] C. I. Shulman, R. W. Smith, and A. Rose, *Phys. Rev.*, **92**, 857 (1953).

TABLE 15. TECHNICAL

| | Selenium[1,2] | Cadmium sulfide[3,4,5,7,20,21] | Cadmium selenide[4,6,21] |
|---|---|---|---|
| Transfer function $R$, $\rho$, or $I = f$ (incident light flux $L$) | Gray Se: $I$ prop $L^{0.5}$ Amorph. Se: $I$ prop $L^{0.9}$ | $I$ proportional $L^n$; $n$, at low light level, 1 to 3; at high level, between 0.8 and 1 | Current increases with light first steeply, later with gradually diminishing slope |
| Luminous sensitivity | Varies widely 100–3,000 $\mu$A/L | | Order of 10 mA/L |
| Condition in darkness | Gray Se, commercial cells, $10^4$–$10^7$ ohms; amorph. Se layers, $10^8$–$10^{10}$ ohms | $2 \times 10^8$ ohm-cm resistivity; powder can be made $10^{10}$–10 ohm-cm; commercial cells resistance $10^9$ ohms | Dark resistance of commercial cells $> 10^{10}$ ohms |
| Condition when illuminated | Gray Se: Light flux of $0.1\ L$ may reduce resistivity by factor of 3 | Resistance at 1 ft-c reduced by factor $10^6$ | Resistance at 1 ft-c $5.10^5$ ohms; resistance reduced from dark by factor of $10^6$ |
| Quantum yield or amplification | About 1 | For X rays: $10^6$ For $\alpha$ or $\beta$ rays: $10^8$ to $10^{10}$ | |
| Voltage-current characteristics | | For crystals essentially linear | Essentially linear |
| Time response | Varies widely; gray Se 50 msec to minutes; amorph. 50 $\mu$sec. Overswing at low voltages | Rise time for 10 ft-c, 30 msec; for 1 ft-c, 300 msec. Decay time for 10 ft-c, 10 msec; for 1 ft-c, 55 msec | 1.5 msec at high light level 15 msec at 1 ft-c sintered layer: 1 msec at 10 ft-c |
| Effect of temperature | Dark current increases at higher temp.; photocurrent increases steeply above 10°C | Sensitivity is minimum at 50°C; increases for decreased as well as increased temp. Temp. hysteresis | Current increases 100% if temp. varies from $+25$ to $-25$°C |
| Noise | | Essentially lower than associated circuit | |
| Fatigue-effect lifetime | | | |

| Lead sulfide[11,12,13,18] | Thallium sulfide[8,9,10] | Silicone[14,19] | Germanium[15,16,17,19] |
|---|---|---|---|
| Essentially linear | Current increases with light first steeply, later with gradually diminishing slope | Essentially linear | Essentially linear |
| 3 mA/L | 1–1.5 mA/L | Order of 60 $\mu$A/L | 30 mA/L at 2400°K color temperature |
| Technical cells $10^5$–$10^7$ ohms | Varies widely $10^4$–$10^8$ ohms, frequently $10^6$–$10^7$ | | Dark current $<20\,\mu$A (ordinarily 1–5 $\mu$A) |
| | Illumination by 0.25 ft-c decreases resistance by factor of 5 | $(I$ photo: $I$ dark$)/L$ $= 12$ to $20$ | Photo current about 300 $\mu$A for 10 mL |
| | At max response, between 15 and 24 | | 1 |
| | | Essentially linear, except at fields above 200 volts/cm | Current rises first with increased voltage, reaches saturation level which depends upon light level |
| 0.1 msec | 1.3 msec; varies widely with illumination and temperature | 0.3 msec | Time constant $<10^{-5}$ sec |
| | Pronounced effect; dark resistance doubles if temperature reduced from 25 to 15°C | | Temp. increase from 20 to 55°C increases dark current by factor of 10, photocurrent by larger factor |
| | Shelf aging increases signal-to-noise ratio | | Short-circuit noise current $2.10^{-11}$ amp in 1 cps band at 1,000 cps |
| | Strong light causes fatigue; UV can cause permanent damage | No noticeable fatigue; can be exposed to high temperature without permanent damage | |

electrodes, which cause an ohmic contact at the interfaces. The noise spectrum flattens off at low frequencies.

Photoconductivity caused by exposure to visible light can be reduced (quenched) by the radiation from a second, infrared light source. Taft and Hebb[1] have shown that the photoconductive current caused in a cadmium sulfide crystal by a radiation of about 550 m$\mu$ can be reduced by additional radiation; the quenching effect is largest for an additional radiation of a wavelength of about 820 to 1,380 m$\mu$. An explanation of the effect is given in the same paper.

The physical properties of the most frequently used photoconductors are summarized in Table 15. It is to be noted that the functional relationships between the parameters for the different photoconductors, as well as their physical properties, may change by orders of magnitude in the presence of impurities or vary with the method of processing. The numerical values found by different authors vary widely, therefore.

Methods for the preparation of photoconductive layers will be found in the reference column on Table 15.

(1) V. K. Zworykin and G. E. Ramberg, "Photoelectricity," p. 181, John Wiley & Sons, Inc., New York, 1950; (2) P. K. Weimer and A. D. Cope, *RCA Rev.*, **12**, 314 (1951); (3) R. Frerichs, *Phys. Rev.*, **72**, 594 (1947); (4) S. M. Thomsen and R. H. Bube, *Rev. Sci. Instr.*, **26**, 664 (1955); (5) R. H. Bube and S. M. Thomsen, *J. Chem. Phys.*, **23**, 15 (1955); (6) F. H. Nicoll and B. Kazan, *J. Opt. Soc. Am.*, **45**, 647 (1955); (7) R. W. Smith and A. Rose, *Phys. Rev.*, **92**(A), 857 (1953); (8) A. Von Hippel et al., *J. Chem. Phys.*, **14**, 370 (1946); (9) A. Von Hippel and E. S. Rittner, *J. Chem. Phys.*, **14**, 370 (1946); (10) C. W. Hewlett, *Gen. Elec. Rev.*, **50**, 22 (April, 1947); (11) R. J. Cashman, *OSRD Rept.* 5998, October, 1945; (12) L. Sosnowski, J. Starkiewicz, and O. Simpson, *Nature*, **159**, 818 (June, 1947); (13) B. Wolfe, *Rev. Sci. Instr.*, **27**, 60 (1956); (14) G. K. Teal, J. R. Fisher, and A. W. Treptow, *J. Appl. Phys.*, **17**, 879 (1946); (15) J. N. Shive, *Bell Lab. Record*, **28**, 337 (1950); (16) S. Benzer, *Phys. Rev.*, **72**, 1267 (1947); (17) J. N. Shive, *Proc. IRE*, **40**, 1410 (1952); (18) R. P. Chasmar, in R. G. Breckenridge, B. R. Russell, and E. E. Hahn (eds.), "Photoconductivity Conference," John Wiley & Sons, Inc., New York, 1956; also G. W. Mahlman, W. B. Nottingham, and J. C. Slater, in R. G. Breckenridge, B. R. Russell, and E. E. Hahn (eds.), "Photoconductivity Conference," John Wiley & Sons, Inc., New York, 1956; (19) R. G. Seed, Germanium Photosensitive Devices, *Bulletin*, Transistor Products, Inc., Waltham, Mass.; (20) *Manual*, RCA Tube Div., Harrison, N.J.; (21) *Pamphlet*, "Clairex Crystal Photocell," Clairex Corporation, New York.

For general references on photoconductivity, see R. G. Breckenridge, B. R. Russell, and E. E. Hahn (eds.), "Photoconductivity Conference (Atlantic City, November, 1954)," John Wiley & Sons, Inc., New York, and Chapman and Hall, Ltd., London, 1956; R. C. Jones, "Performance of Visible and Infrared

---

[1] E. A. Taft and M. H. Hebb, *J. Opt. Soc. Am.*, **42**, 249 (1952).

Detectors," in "Advances in Electronics," p. 63, Academic Press, Inc., New York, 1953. For an extensive list of references, see also P. Görlich, *Arch. tech. Messen*, J 394–1, September, 1958.

### 5-13. PHOTOVOLTAIC CELLS

The photovoltaic, or barrier-layer, cell, also Photronic cell (trade name, Weston Instruments Division of Daystrom, Inc.), consists basically of three solid layers, Fig. (5-1)30, a base plate $B$, a semiconducting layer $S$, and a transparent metal electrode $T$. Proper processing, produces a thin insulating barrier layer between the semiconductor and the transparent metal layer. If light strikes the barrier layer, a voltage is generated between the output terminals, the base plate being positive, the transparent electrode negative.

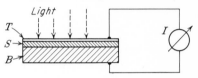

Crystalline selenium is mainly used as a semiconducting material. Cells with silicon and germanium layers have been described lately. The cuprous oxide employed for-

FIG. (5-1)30. Photogalvanic system: $B$, base plate; $S$, layer forming barrier layer at interface with transparent metal layer $T$.

merly is rarely used today. Cuprous oxide lends itself to the production of two different cell types: one where the barrier layer is formed between the base plate and the cuprous oxide layer (back-wall cells) and another where the barrier layer is formed between the cuprous oxide and the transparent metal layer (front-wall cells).

For references concerning processing of photogalvanic cells or semiconducting substances other than those mentioned above, see the review articles by Zworykin and Ramberg, *op. cit.*, chap. 11 and references, pp. 214–215; D. Geist, *Arch. tech. Messen*, J 392–2, December, 1954; A. Kellermann, *Arch. tech. Messen*, J 392–3, February, 1955.

The physical mechanism which initiates the generation of an emf in photogalvanic cells is the absorption of an incident quantum which leads to the formation of an electron and hole pair. The electric field in the barrier layer causes a separation of the carriers and, in further sequence, the generation of a charge between the semiconductor and the metal layer. The charges can either flow back through the barrier layer (internal current) or can cause a current in an external circuit connected to the terminals.

The open-circuit voltage is independent of the area of the photogalvanic cells. It increases in a near-logarithmic fashion with increasing illumination, as shown in Fig. (5-1)31. The short-circuit current is proportional to the area of the cell and increases linearly with the level of illumination, as shown in Fig. (5-1)32. The transfer function (output current versus illumination) deviates from linearity (1) at very high levels of illumination, (2) if only a part of the surface is illuminated, and (3) if the load resistance is not small compared to the internal resistance of the cell, as shown in Fig. (5-1)32.

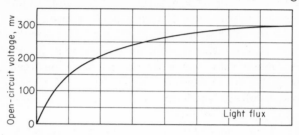

FIG. (5-1)31. Open-circuit output voltage as a function of the incident light flux for a photogalvanic cell.

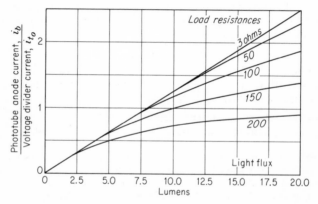

FIG. (5-1)32. Output current as a function of the incident light flux for a photogalvanic cell; parameter load resistance.

The variation of the characteristic with the load resistance may be considered an advantage; it enables the user to construct instruments with either linear calibration curves or curves that approach a logarithmic function and cover several orders of magnitude of light flux without switching the range of the meter (exposure meters).

Linearity of the transfer characteristic can be obtained by compensation methods devised by N. R. Campbell and M. K. Freeth, *J. Sci. Instr.*, **11**, 125 (1934); by R. A. Houstoun and A. F. Howatson, *Phil. Mag.*, **36**, 279 (1945); by L. A. Wood, *Rev. Sci. Instr.*, **7**, 157 (1936); or by the feedback circuit described by E. S. Rittner, *Rev. Sci. Instr.*, **18**, 36 (1947).

The sensitivity of photogalvanic cells can be fairly high; that of commercial selenium cells is of the order of 1 mA/lumen for small load resistances. The open-circuit output voltage of selenium cells, at high levels of illumination, can attain values from 200 to 600 mV. The sensitivity of cuprous oxide cells is only about one-tenth that of selenium cells. The power output is about $10^{-8}$ watt/cm$^2$ at a light flux of 0.01 lumen and increases to $10^{-5}$ watt/cm$^2$ at 10 lumens. In the same light-flux interval the internal resistance decreases from $10^5$ to about $10^3$ ohms/cm$^2$ of cell area.

The spectral response of selenium cells extends from about 250 to 750 m$\mu$, with a maximum response at about 570 m$\mu$; occasionally values for the threshold wavelength of 850 to 1,200 m$\mu$ and a maximum response at 720 m$\mu$ are found.[1] Selenium cells respond also to X rays.[2] The spectral response of the cuprous oxide front-wall cell extends from less than 300 to 630 m$\mu$ with a maximum at 520 m$\mu$; that of the cuprous oxide back-wall cell from 570 to 1,400 m$\mu$ with a maximum response near 630 m$\mu$. The response of germanium cells is primarily in the infrared part of the spectrum (maximum response 1,500 m$\mu$, threshold about 1,700 m$\mu$) but extends into the visible region. The threshold wavelength of silicon is 1,200 m$\mu$. The quantum yield in photogalvanic cells is in the vicinity of 1.

The electrical (voltage-current) characteristic of the photogal-vanic cell is complicated by the nonlinearity of the system. The internal resistance, i.e., the resistance of the barrier layer, depends upon the current density and therefore upon the illumination and the load resistance. Hence, impedance matching is possible only under a given level of illumination. A voltage-current characteristic (relative values only) is shown in Fig. (5-1)33. The base curve $A$ denotes the rectifier characteristic of the system without illumina-tion. By plotting the resistance-load line, the output voltage, current, and power as well as the internal resistance can be found in the usual way.

The generation of an emf within the cell takes place a short time after the exposure to light (of the order of $10^{-6}$ sec). However, the build-up of the output voltage is delayed by the large internal capacitance of the barrier-layer system. The equivalent circuit of a photogalvanic cell is shown in Fig. (5-1)34. The capacitance of selenium cells is very high, of the order of 0.01 to 0.1 farad/cm$^2$, and is less for cuprous oxide cells. Germanium cells have a much faster

[1] Kellermann, *loc. cit.*

[2] B. Lange and P. Selény, *Naturwissenschaften*, **19**, 639 (1931); K. Scharf and O. Weinbaum, *Z. Physik*, **80**, 465 (1933).

response and have been successfully operated with light modulated at a frequency of several megacycles. The decay time of the output voltage of germanium cells is of the order of 10 $\mu$sec and increases strongly at lower temperature.[1]

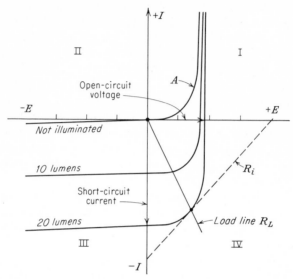

Fig. (5-1)33. Voltage-current characteristic of a photo-galvanic cell.

The influence of the temperature upon the open-circuit voltage is considerable; the voltage decreases rapidly with increased temperature. The influence of the temperature upon the short-circuit current is much less pronounced. Figure (5-1)35 shows the variation with temperature of the current output for different load conditions. Temporary (reversible) fatigue and the permanent reduction of sensitivity in photogalvanic cells have been observed by some authors, while others have experienced a variation of less than 1 per cent in the course of a year.

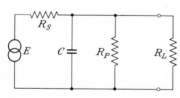

Fig. (5-1)34. Equivalent circuit of a photogalvanic cell: $R_S$, series resistance; $C$, capacitance; $R_P$, parallel resistance of the cell; $R_L$, load resistance.

The application of an auxiliary voltage and a load resistance in series with the photosensitive barrier layer changes the operation

[1] M. Becker and H. Y. Fan, *Phys. Rev.*, **78**, 301 (1950).

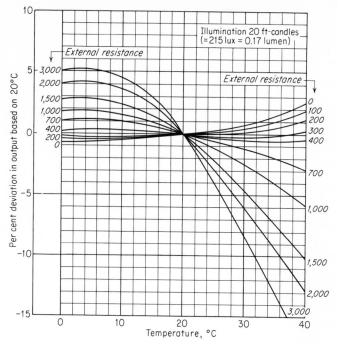

FIG. (5-1)35. Temperature effect on a photogalvanic cell. Ordinate, variation of current output in percentage of the output at 20°C; parameter load resistance (*Weston Instruments, Division of Daystrom, Inc.; by permission*).

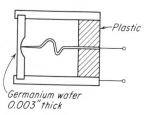

FIG. (5-1)36. Germanium photodiode, cross section [*from J. N. Shive, Bell Labs. Record*, **28**, 337 (1950); *by permission*].

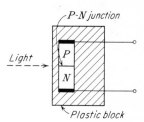

FIG. (5-1)37. *PN* junction photodiode [*from J. N. Shive, Proc. IRE*, **40**, 1410, (1952); *by permission*].

from that of a photogalvanic cell to that of a photodiode. Photodiodes have in general a much smaller area; therefore, their capacitance is smaller and their frequency response is higher. The construction of two types of germanium photodiodes is shown schematically in Figs. (5-1)36 and (5-1)37. The sensitive area is usually of the order of 1 mm². The operation of the photodiode can be understood from the characteristic of Fig. (5-1)33. Without illumination the barrier layer acts as a rectifier. The forward current (first quadrant) rises sharply with the applied voltage $+E$ in the forward direction. The influence of the illumination is small, since the forward dark current is high already. The dark current in the reverse direction $-E$ is small, but the current in the reverse direction increases strongly with increased level of illumination. The photodiode is operated in this region of voltages and currents, i.e., in the third quadrant.

Figures for sensitivity of photodiodes quoted in the literature vary from less than 10 to about 50 mA/lumen. The spectral response extends from the visible to the infrared region, about 2.2 $\mu$, with a maximum relative sensitivity in the vicinity of 1.5 $\mu$. The output noise level is around 3 to 10 $\mu$V and exceeds that of photovoltaic cells; the noise for 1 cps band width decreases with increased frequency. Increase of temperature causes an increase of the photocurrent as well as of the dark current. The use of an a-c supply voltage for the operation of barrier-layer cells has been described by Sargrove.[1]

The combination of two photodiodes, back to back, forms a phototransistor. It can be connected in any one of the customary circuits for $npn$ transistors, the output current being controlled by light falling into the central $p$ region. Both the photocurrent and the dark current appear amplified at the transistor output.

## 5-14. THERMAL-RADIATION TRANSDUCERS

Thermal-radiation transducers are primarily applicable to the measurement of infrared radiation, where the energy of the single quantum is too small to cause photoelectric effects. All thermal-radiation transducers consist essentially of a thin metal strip which absorbs radiation, so that its temperature is increased. The temperature is measured either (a) with a thermoelement, (b) through the accompanying variation of resistance with temperature (bolometers), or (c) by increase of thermionic emission from a filament.

a. *Radiation Thermoelements.* The radiation thermoelement shown schematically in Fig. (5-1)38 consists of a thin disk $F$ on which a thermoelement $T$ is fastened. Radiation striking the disk is absorbed and causes an increase of the receiver temperature. The system comes to equilibrium when the power dissipated from the

[1] J. A. Sargrove, *J. Brit. Inst. Radio Engrs.*, **7**, 86 (May–June, 1947).

disk by radiation, conduction, and convection is equal to the incident radiation power. For most practical applications, the output voltage is a linear function of the incident radiation power density in microwatts per square centimeter.

The receiver disk is usually made from a thin sheet of gold leaf, silver, or platinum foil, and the thermocouple wires are soldered or

Fig. (5-1)38. Radiation thermoelement: $F$, disk; $T$, thermoelement.

Fig. (5-1)39. Radiation thermoelement: $T$, thermojunction.

welded to the foil (e.g., by a capacitor discharge). A different construction is shown in Fig. (5-1)39; it consists of a flat strip of two different metals forming a thermoelement.

Numerous radiation-thermoelement constructions have been developed, primarily with the aim to increase the sensitivity and shorten the response time of the element. The following means have been employed:

1. USE OF VERY THIN THERMOELEMENTS. Thin elements have a small heat capacity and, therefore, a fast response; also the heat conduction from the thermojunction through the thermoelement wires or foils diminishes with a decrease of their cross section. Moll and Burger[1] have soldered together relatively heavy sheets of the two metals which form the thermojunction. The resulting sheet was then rolled down to a foil of 5 $\mu$ thickness, from which strips were cut. Another construction uses a thin insulating carrier, on which, on slightly overlapping adjacent areas, two different metals are deposited by cathode sputtering or evaporating.[2] The elements produced in this way have high mechanical stability, and the method permits a wide choice of metals. Similar radiation thermoelements (0.1 $\mu$ thickness) can be made by electrolytic deposition.[3]

2. MULTIPLE THERMOELEMENTS. The output voltage can be increased by the use of multiple thermocouples, all connected in series

[1] W. J. H. Moll and H. C. Burger, *Z. Physik*, **32**, 575 (1925).

[2] H. C. Burger and P. H. van Cittert, *Z. Physik*, **66**, 210 (1930); L. Harris and E. A. Johnson, *Rev. Sci. Instr.*, **4**, 454 (1933), and 5, 153 (1934).

[3] C. Müller, *Naturwissenschaften*, **19**, 416 (1931), and R. V. Jones, *J. Sci. Instr.*, **14**, 83 (1937).

(thermopiles), as shown in Fig. (5-1)40. One set of thermojunctions is exposed to the incident radiation, the other side is protected from it. A series-parallel linear thermopile has been described by Crane and Blacet.[1] Whether or not the use of thermopiles offers an advantage over the single thermoelement depends upon the level of the incident radiation, its cross-sectional area, and the input impedance of the subsequent stage.[2] Roess and Dacus[3] have shown that the signal-to-noise ratio of the radiation thermoelement is a maximum for the single thermoelement.

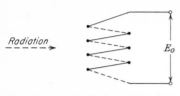

*Radiation*

$E_o$

FIG. (5-1)40. Radiation thermopile made from several thermoelements, alternate junctions exposed to radiation.

3. REDUCTION OF CONVECTION AND CONDUCTION TO ATMOSPHERE SURROUNDING THE THERMOELEMENT. These losses can be reduced by enclosing the thermojunction in an evacuated envelope. The sensitivity rises in the form of an S-shaped curve, starting at about 1 mm Hg, and reaches a final value at about $10^{-4}$ mm Hg. The increase of sensitivity through evacuation depends upon the construction of the element and amounts to a factor varying between 10 and several hundred.[4] Of course, the enclosure envelope causes absorption of radiation in the glass wall and limits the spectral range for which the thermoelement can be used. This difficulty can be remedied, to some extent, by the use of thin windows or windows of quartz and other infrared or ultraviolet transparent materials.

4. REDUCTION OF RADIATION LOSSES FROM THERMOJUNCTION. Since these losses depend upon the temperature of the junction, an operation of the element at low temperatures can increase the sensitivity of the system. Cartwright,[5] who used the radiation thermoelement at the temperature of liquid air, found an increase of sensitivity by a factor of 10 as compared to operation at room temperature. (For other possible advantages of operation of the element at low temperature, see the paper by Cartwright.)

Different methods have been used for blackening the exposed surface in order to absorb as much as possible of the incident radiation. Soot, turpentine smoke, and camphor smoke are useful in air

[1] R. A. Crane and F. E. Blacet, *Rev. Sci. Instr.*, **21**, 259 (1950).

[2] See F. Kerkhof, *Arch. tech. Messen*, J 2404–1, October, 1940.

[3] L. C. Roess and E. N. Dacus, *Rev. Sci. Instr.*, **16**, 164 (1945).

[4] G. Rosenthal, *Z. Instrumentenk.*, **59**, 432 and 457 (1939).

[5] C. H. Cartwright, *Rev. Sci. Instr.*, **4**, 382 (1938).

but not *in vacuo* because of their low thermal conductivity. Colloidal graphite (Aquadag) has been recommended. Platinum black, electrolytically deposited from a solution of 1 part platinum chloride and 0.008 part of lead acetate in 30 parts of water at a current density of 10 mA/cm², furnishes good results. Blackening by evaporation of bismuth and antimony in the presence of air at a pressure of 0.5 to several millimeters Hg has been reported by Pfund[1] and by Burger and van Cittert;[2] evaporation of gold in an $H_2$ atmosphere by Roess and Dacus;[3] and evaporation of aluminum by Gilham.[4]

With proper blackening the radiation thermoelement has an equal spectral response over the entire spectrum from the ultraviolet to the far infrared. The only limitation is that imposed by the window material mentioned above.

The sensitivity of radiation thermoelements is of the order of a fraction of a microvolt for an incident radiation of 1 $\mu$W/cm². Data furnished by different authors frequently do not permit comparisons. The useful receiving areas of elements vary from a fraction of a square millimeter to several hundred square millimeters. For optimum sensitivity, the receiving area should not exceed the cross section of the incident beam. The output impedance is between 10 and 100 ohms and can reach higher values with multiple thermoelements. The response time of radiation thermoelements may vary from a fraction of a second to several seconds. Systems with very short response time ($\sim$0.1 sec) are described by Roess and Dacus,[5] as well as by Cary and George.[6]

For references and bibliography, see the review papers by M. Czerny and H. Röder, *Ergeb. exact. Naturw.*, **27**, 70 (1938); also H. Cary and K. P. George, *loc. cit.*; F. Kerkhof, *loc. cit.*; L. Geiling, *Arch. tech. Messen*, J 2404–2, November, 1955, and J 2404–3, February, 1956. An extensive study of the ultimate sensitivity is due to D. F. Horning and B. J. O'Keefe, *Rev. Sci. Instr.*, **18**, 474 (1947). A further study, primarily concerned with noise in thermal-radiation receivers, is published by R. C. Jones, "Performance of Detectors for Visible and Infrared Radiation," in L. Marton (ed.), "Advances in Electronics," vol. 5, p. 1, Academic Press, Inc., New York, 1953.

[1] A. H. Pfund, *Rev. Sci. Instr.*, **8**, 417 (1937).
[2] H. C. Burger and P. H. van Cittert, *Z. Physik*, **66**, 656 (1933).
[3] Roess and Dacus, *loc. cit.*
[4] E. J. Gilham, *J. Sci. Instr.*, **33**, 338 (1956). For further information on blackening methods, see A. H. Pfund, in W. E. Forsythe (ed.), "Measurement of Radiant Energy," McGraw-Hill Book Company, Inc., chap. 6, pp. 210ff., New York, 1937.
[5] Roess and Dacus, *loc. cit.*
[6] H. Cary and K. P. George, *Phys. Rev.*, **71**, 276 (1947).

*b. Radiation Bolometers.*[1] The radiation bolometer, Fig. (5-1)41, consists essentially of a thin strip of a material which absorbs radiation; the temperature increase causes a variation of the electrical resistance of the strip.

The bolometer strip is commonly used in a Wheatstone bridge or a similar suitable network. The bridge is balanced while the bolometer is protected from radiation. The ratio of voltage unbalance caused by exposure of the bolometer strip to the unit radiation (in watts per square centimeter) is frequently used as a measure for the sensitivity of the bolometer.[2]

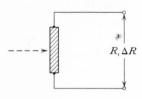

$R, \Delta R$

The transfer characteristics of metal bolometers are essentially linear, at least within the generally employed practical limits. The sensitivity is proportional to the current passing through the bolometer at low current densities; as the current density increases, the sensitivity reaches a maximum, since the radiation losses increase proportionally with $T^3$.

FIG. (5-1)41. Radiation bolometer.

Design considerations for bolometers of high sensitivity and short time response are analogous to those for radiation thermocouples treated above. The sensitivity can be increased by reduction of the thermal losses, the speed of response by reduction of the heat capacity of the strip. The bolometer is frequently made from a very thin strip of metal, such as Wollaston foil (platinum), or evaporated on a nonconducting carrier (nitrocellulose pellicles). The reduction of the thickness cannot be driven too far, since the resistance-temperature coefficient of very thin metal layers is considerably less than that of metal in bulk form.

A number of older constructions are reviewed by T. Dreisch, in H. Geiger–K. Scheel (eds.), "Handbuch der Physik," chap. 26, pp. 842ff., 1928. Recent work on metal bolometers may be found by E. B. Baker and C. D. Robb, *Rev. Sci. Instr.*, **14**, 356 (1943); I. Amdur and C. F. Glick, *Rev. Sci. Instr.*, **16**, 117 (1945); W. G. Langton, *J. Opt. Soc. Am.*, **36**, 355 (1946); C. B. Aiken, W. H. Carter, Jr., and F. S. Philips, *Rev. Sci. Instr.*, **17**, 377 (1946); B. H. Billings, E. E. Barr, and W. L. Hyde, *J. Opt. Soc. Am.*, **36**, 354 (1946), and **37**, 123 (1947), and *Rev. Sci. Instr.*, **18**, 429 (1947). A theory of the metal bolometer may be found by R. P. Chasmar, W. H. Mitchell, and A. Rennie, *J. Opt. Soc. Am.*, **46**,

---

[1] The bolometer is also used for the measurement of physical phenomena other than radiation, e.g., mechanical displacements or alternating currents. Such applications are not considered in this section.

[2] P. Moon and W. R. Mills, Jr., *Rev. Sci. Instr.*, **6**, 8 (1935).

469 (1956); a theoretical study primarily concerning the frequency characteristic of bolometers exposed to radiation of a periodically fluctuating intensity is due to J. N. Shive, *J. Appl. Physics*, **18**, 389 (1947).

Dielectric materials, instead of metals, can be used for the construction of bolometers. The first dielectric bolometer, using cellophane or pliofilm (in an a-c bridge), is described by Moon and Steinhard.[1] A theoretical study by Ewles[2] reveals that the sensitivities of radiation thermoelements and metallic radiation bolometers are almost alike, while that of dielectric bolometers can be considerably higher (see, however, below).

The resistance-temperature coefficient of semiconductors can be much higher than that of metals (by a factor of about 10), and a number of authors have used semiconductors for the construction of bolometers. Bauer[3] has investigated bolometers made from cuprous oxide; Brattain,[4] Dodd,[5] and Wormser[6] have employed thermistors. Gilham[7] has built a bolometer from semiconducting antimony which has a high resistance (18 M$\Omega$). Ward[8] described a bolometer for high radiation intensities, 1.4 cal/(cm$^2$)(min), containing one thermistor exposed to radiation and another measuring environmental temperature. The indication of this system is almost independent of the environmental temperature in the range between $-19$ and $+0.5°$C.

The resistivity of electronic conductors becomes zero at very low temperatures (superconduction). In the transition zone from normal conduction to superconduction the resistance-temperature coefficient attains very high values (of the order of several thousand per cent per degree), so that bolometers operated in the transition range are very sensitive. The operation at low temperature also offers other advantages, such as the reduction of noise and of thermal reradiation. Superconducting bolometers were first built with tantalum wires operated between 3.22 and 3.23°K.[9] Later constructions used columbium nitride operated at 15°K.[10] Superconducting

[1] P. Moon and L. R. Steinhard, *J. Opt. Soc. Am.*, **28**, 148 (1938).

[2] J. Ewles, *J. Sci. Instr.*, **24**, 57 (1947).

[3] G. Bauer, *Phys. Z.*, **44**, 53 (1943).

[4] W. H. Brattain, *J. Opt. Soc. Am.*, **36**, 354 (1946).

[5] R. E. Dodd, *J. Sci. Instr.*, **28**, 12 (1951).

[6] E. M. Wormser, *J. Opt. Soc. Am.*, **43**, 15 (1953).

[7] E. J. Gilham, *loc. cit.*

[8] W. H. Ward, *J. Sci. Instr.*, **34**, 317 (1957).

[9] D. H. Andrews et al., *Rev. Sci. Instr.*, **13**, 281 (1942).

[10] D. H. Andrews, R. M. Milton, and W. DeSorbo, *J. Opt. Soc. Am.*, **36**, 353 (1946), and N. Fuson, *J. Opt. Soc. Am.*, **38**, 845 (1948).

bolometers are "at best 2 to 3 times better than the Golay pneumatic detector[1] and 10 to 20 times better than the best thermopiles and room-temperature bolometers."[2]

Comparison of different bolometers is very difficult. A considerable number of parameters enter in the performance characteristics of bolometers as they do in other light transducers, as has been pointed out by Jones,[3] so that data on the sensitivity or the input or output characteristics have very limited value and are omitted here. Comparative studies made by Fuson[4] are tabulated in Table 16; others made by Billings, Barr, and Hyde[5] in Table 17. The

TABLE 16.  COMPARISON OF THERMAL-RADIATION RECEIVERS*

| Type | Shortest response time, msec | $\Delta J_0$, watts† |
|------|------------------------------|----------------------|
| Superconducting bolometer | 0.3 | $3.4 \times 10^{-10}$ |
| Platinum strip bolometer | 4.1 | $7.2 \times 10^{-10}$ |
| Evaporated nickel bolometer | 4.0 | $209.0 \times 10^{-10}$ |
| Semiconducting bolometer | 3.0 | $167.0 \times 10^{-10}$ |
| Thermocouple | 36.0 | $1.16 \times 10^{-10}$ |
| Golay (pneumatic) detector | 5.0 | $3.3 \times 10^{-10}$ |

\* N. Fuson, *J. Opt. Soc. Am.*, **38**, 845 (1948).

† $\Delta J_0$ is that value of steady incident radiation power which produces a steady output equal to the rms Johnson noise.

TABLE 17.  COMPARISON OF THERMAL RECEIVERS*

| Type | Response time | Calc. threshold, watts |
|------|---------------|------------------------|
| Metal bolometer | 4.0 msec | $3.3 \times 10^{-8}$ |
| Thermistor bolometer | 3.0 msec | $7.2 \times 10^{-8}$ |
| Dielectric bolometer | 3.5 sec | $2.3 \times 10^{-4}$ |
| Thermocouple | 0.2 sec | $3.0 \times 10^{-6}$ |

\* B. H. Billings, E. E. Barr, and W. L. Hyde, *J. Opt. Soc. Am.*, **37**, 123 (1947).

comparison in the originals of these tables is made under certain standardized conditions. Variations of these conditions may change the relative merit of these bolometers by orders of magnitudes.

[1] M. J. E. Golay, *Rev. Sci. Instr.*, **18**, 357 (1947).

[2] N. Fuson, *loc. cit.*

[3] R. C. Jones, *Rev. Sci. Instr.*, **24**, 1035 (1953).

[4] Fuson, *loc. cit.*

[5] B. H. Billings, E. E. Barr, and W. L. Hyde, *J. Opt. Soc. Am.*, **37**, 123 (1947).

*c. Thermionic Detector.* The system is similar to a thermionic diode with a directly heated cathode, as shown in Fig. (5-1)42. The applied plate voltage and the filament temperature are chosen so that operation in the temperature-limited region occurs. If radiation strikes the cathode, its temperature will be raised, and the plate current will increase. For theory and references, see Jones.[1]

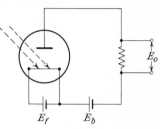

FIG. (5-1)42. Thermionic radiation detector: $E_f$, filament voltage source; $E_b$, plate voltage.

## 5-2. Transducers for Ionizing Radiation

The transducers described in this section are based upon the ion-producing effect of radiation; they include the ionization chamber (5-21) and the gas-filled counters (5-22), as well as the crystal counter(5-23). Scintillation counters and Cerenkov counters are omitted, since they are not considered electrical instrumentation *elements*, but rather combinations of systems that convert ionizing radiation into light with transducers that convert light into electric signals.

Ionizing radiation is either of an electromagnetic nature, such as X rays or gamma rays, or of a corpuscular nature, such as alpha or beta rays.

Ionizing radiation can sometimes be converted into electric signals by transducers for optical radiation. Photomultipliers, for instance, can be sensitive for gamma radiation;[2] also the response of some photoconductive and photogalvanic systems extends into the X-ray and gamma-ray region, and some of them respond to alpha and beta radiation (see 5-12 and 5-13). Radiation of charged particles can be converted into electric signals also by transducers for free charges (4-1).

Ionization chambers and proportional counters (as well as scintillation counters) measure the energy per cubic centimeter and per second of the absorbed radiation, while Geiger counters measure the number of particles without regard to their energy.

### 5-21. Ionization Chambers

The ionization chamber, schematically illustrated in Fig. (5-2)1, consists essentially of two electrodes, *A* and *B*, well insulated from

[1] R. C. Jones, in L. Marton (ed.), "Advances in Electronics," sec. IV, 4, p. 50, Academic Press, Inc., New York, 1953.

[2] J. D. Graves and G. E. Koch, *Rev. Sci. Instr.*, **21**, 304 (1950).

each other within a chamber $C$. A voltage source $E$ produces an electric field between the electrodes. Ionizing radiation entering the space between the electrodes causes the removal of electrons from some of the gas molecules, which thereby become positive ions. The electrons and the ions are separated by the applied electric

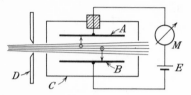

FIG. (5-2)1. Ionization chamber: $A$ and $B$, electrodes; $C$, chamber; $D$, diaphragm; $E$, voltage source; $M$, meter.

field; they move in opposite direction toward the electrodes and in doing so form an electric current, which is measured by a meter $M$.

On their way to the electrodes, the ions and electrons may recombine to neutral molecules. If the applied voltage is low, the velocity of the electrons and ions is small, and the electrons and ions (or negative and positive ions) will have time to recombine; the current is small. The current increases with the applied voltage, Fig. (5-2)2, and reaches a constant value (satu-ration current) when all ions are col-lected on the electrodes before they have a chance to recombine.

The field strength required to reach saturation rises with the amount of ionization, i.e., with incident radiation intensity, and with the pressure in the chamber. Higher field strengths are also needed if the ionization is caused by alpha particles (because of the increased recombination in col-umns). If electron attachment takes place in contaminated gases, satu-ration cannot be reached, and the

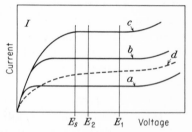

FIG. (5-2)2. Voltage–current charac-teristic of an ionization chamber: curves $a$, $b$, $c$ obtained with increasing amount of ionization; $d$, curve ob-served in the presence of electron attachment; $E_s$, voltage at which saturation occurs; operating range between $E_1$ and $E_2$.

current may increase steadily with increased voltage, as shown in Fig. (5-2)2, curve $d$.

At very high ionization density or in high-pressure chambers where the voltages required to obtain saturation are impractically high, operation in the range of the very beginning of the voltage-current characteristic is recommended (at about 1 per cent of the saturation voltage). In this region the current is proportional to the square root of the incident radiation intensity.

For a steady flux of ionizing radiation, and if the losses of charges within the chamber can be neglected, the saturation current is

$$I = N_0 Se$$

where $N_0$ is the number of particles arriving in the sensitive volume of the chamber per second and causing ionization, $S$ is the number of ion pairs formed per ionizing particle, and $e$ the electron charge. The current can be measured with a galvanometer, but is in general too small for this instrument. Measurement of the voltage drop across a large resistance or electrometric methods are usually employed. A convenient way to measure very small radiation intensities consists in charging the ionization chamber to a voltage $E_1$, Fig. (5-2)2, and exposing it to radiation for a time $t$. The chamber will discharge to a voltage $E_2$ which should still be above $E_s$, the voltage at which saturation occurs. If $E_1$ and $E_2$ are measured electrometrically, i.e., without drawing current from the charged chamber, the average current during the time $t$ can be found from

$$I = \frac{C(E_1 - E_2)}{t}$$

where $C$ is the capacitance of the ionization chamber (integrating-chamber method).

The sensitivity of the ionization chamber increases with the number of ion pairs formed per incident ionizing particle and decreases with the number of ions lost during their travel through the chamber.

The number of ion pairs formed per incident ionizing particle is, approximately,

$$n = Slp \tag{1}$$

where $S$ is the specific ionization (number of ions per centimeter of length and per atmosphere of pressure), $l$ the average path length of the ionizing particles through the chamber, and $p$ the pressure in atmospheres.

The proportionality of the current with pressure as expressed by Eq. (1) is correct only in the range below 10 atm. At higher pressure the sensitivity increases at a reduced rate, and in the range above 20 to 200 atm (depending upon the field strength) the sensitivity of the chamber decreases, because the mobilities of the ions are reduced and recombination is increased.[1]

[1] H. A. Erickson, *Phys. Rev.*, **27**, 473 (1908).

Losses occur through recombinations. The rate of recombination of positive and negative ions is by a factor of $10^4$ larger than that of positive ions and electrons. Negative ions are formed by electron attachment, primarily in the presence of impurities. Gases of very high purity ($>99$ per cent) should be used to reduce attachment.

RESPONSE TO IONIZATION VARYING WITH TIME. The current passing through an ionization chamber is frequently measured by the voltage drop across a series resistor $R$, as shown in Fig. (5-2)3. The time constant of the ionization-chamber circuit is then determined by the product $RC$. If the resolving time of the subsequent detecting equipment together with the $RC$ network is $\tau$, the value of $\tau$ must be substituted for $RC$.

The capacitor is charged to the voltage $E$. Any transition of charges $dq$ within the chamber causes a discharge of the capacitor and a variation of the voltage across the chamber or across the resistance $R$ of a value

$$\Delta E = \frac{dq}{e} = 1.6 \times 10^{-7} \frac{n}{C}$$

where $C$ is the capacitance of the chamber plus that of the amplifier input, and $n$ is the number of ion pairs formed. The capacitor is recharged from the source $E$ through the resistor $R$. Two cases can be distinguished:

1. Ion pulse chamber. The time constant $RC$ is large compared to the time required to collect the ions. The output pulse rises rapidly at first until all electrons are collected (of the order of 1 $\mu$sec) and then rises slowly until all ions are collected (about 1 msec). The pulse then decays with the time constant $RC$.

The maximum voltage of the pulse in this case is

$$\Delta E_p = \frac{Q_0^+ + Q_0^-}{C} = \frac{N_0 e}{C}$$

where $Q_0^+$ and $Q_0^-$ are the positive and negative charges produced by an ionizing particle, and $C$ is the capacitance of the chamber plus that of the input to the amplifier. $N_0$ is the total number of ions produced in the chamber.

The ion pulse chamber thus furnishes information as to the total number of ions produced in the chamber, but its response is slow, so that the chamber is not suitable for fast counting. The response characteristic of the subsequent amplifier must extend to low frequencies, and the amplifier is, therefore, subject to microphonics of the counter (see below) and a-c pickup.

2. Electron pulse chamber. The time constant $RC$ is larger than the time required to collect the electrons but small compared to the collection time for the ions (order of $RC$ is 10 to 20 $\mu$sec). The pulse height in this case is

$$E_p = \frac{Q_0^-}{C}$$

Since the total number of negative charges is close to that of positive charges, the electron pulse chamber also furnishes information as to the total number of ions produced in the chamber. This

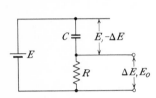

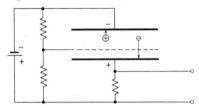

FIG. (5-2)3. Equivalent circuit of an ionization chamber.

FIG. (5-2)4. Ionization chamber with built-in grid.

chamber is faster, but has the disadvantage that the pulse height depends upon the location within the counter where the charges are produced. By the insertion of a grid between the two electrodes, as in the Frish chamber, Fig. (5-2)4, this difficulty can be overcome. The grid is held at a potential between those of the electrodes. Electrons produced between the negative electrode and the grid will pass through the grid and will be recorded; the influence of the ion movement upon the collector is shielded by the grid.

The energy of an ionizing particle that spends its whole energy within the chamber can be found from

$$P = \frac{Q}{e} W_0$$

where $Q$ is the charge induced in the collector, and $W_0$ is the energy required for the formation of an ion pair (about 35 eV for alpha rays in air).

The mean square fluctuation of the output voltage is approximately (within a factor of 2)

$$\overline{\Delta E^2} = \frac{E_o^2 me}{I} = R^2 I \frac{me}{\tau}$$

where $E_o$ is the output voltage, $I$ the current through the resistor $R$, $m$ the average number of ion pairs produced by each particle in the chamber, $e$ the electron charge, and $\tau$ the resolving time of the detecting equipment.

The smallest amount of ionization (instantaneous value) that can be detected with the ionization chamber depends upon the current sensitivity of the stage following the chamber. One roentgen per hour causes an ion current of the order of $10^{-13}$ amp/cm$^3$ at 1 atm. With a 1,000-cm$^3$ chamber, the current is, therefore, $10^{-10}$ amp. If $10^{-16}$ amp can be measured (this is about the practical limit of most electrometrical current measuring systems), 1 $\mu$r/hr can be detected. The integrating method (which measures the total X-ray dose or the average intensity over a long time, rather than instantaneous values of radiation intensity) is, of course, capable of measuring smaller values of radiation intensity.

Errors in ionization-chamber measurements are likely to arise if either the primary particles or the secondary electrons strike the walls or the electrodes. Limiting the cross section of the incident beam, Fig. (5-2)1, diaphragm $D$, and the use of large chambers help to prevent such errors.

The ion current in a small chamber is primarily caused through secondary electron emission from the walls of the chamber. If the chamber is to be used for the absolute measurement of X-ray intensity (which is based on ionization in air), the walls of the chamber should be made of a material which causes per incident X ray the same ionization current as air (air-equivalent chamber).[1] Such materials consist usually of a mixture of graphite, aluminum, silicon, or magnesium with a plastic binder.[2]

Background ionization and spurious discharges may cause further errors and may reduce the ultimate sensitivity of the chamber. The chamber walls are frequently contaminated to a small degree. Alpha particles usually come from contaminations of the chamber walls. Some authors recommend zinc, iron, steel, and aluminum to be used as construction material for the walls and advise against the use of copper, brass, or platinum. A thin layer of carbon black or collodion applied to the inside walls of the ionization chamber frequently reduces the effect of alpha rays. The effect of alpha emission from the wall is reduced if the counter is filled with gas at high pressure. Beta or gamma rays from the outside can be shielded by thick layers of lead or similar materials.

---

[1] H. Fricke and O. Glasser, *Fortschr. Röntgen*, **33**, 239 (1925).

[2] R. Glocker and E. Kaupp, *Strahlentherapie*, **23**, 447 (1926); see also E. Miehlnickel, *Ann. Physik*, **20**, 737 (1934).

Insulation difficulties can be reduced by a guard-ring arrangement, as shown in Fig. (5-2)5. The guard ring frequently serves as an electrostatic shield of the collector; adequate shielding of the collector and the leads should be provided. The charged ionization chamber, in particular the parallel-plate chamber, is in its construction very similar to a condenser microphone. A small displacement

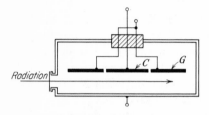

FIG. (5-2)5. Ionization chamber with electrode *C* and guard electrode *G*.

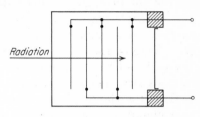

FIG. (5-2)6. Multiple-plate ionization chamber.

FIG. (5-2)7. Thimble ionization chamber.

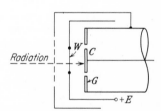

FIG. (5-2)8. Ionization chamber for alpha particles.

of one of the electrodes causes a large current. Counters of this type must be protected from microphonic influences.

Different constructions of ionization chambers are shown in Figs. (5-2)5 to (5-2)8. The parallel-plate chamber, Fig. (5-2)5, has the advantage of a uniform field, so that saturation is easily obtained. The sensitive volume of the chamber, i.e., the volume where an ion pair formed by ionization is recorded, is defined by the cross-sectional area of the incident beam and the length of the collector *C*. Since the field at the edges of a simple collector is not well defined, the collector is frequently surrounded by a guard ring *G* which is kept at or close to the potential of the collector. Ionization takes place primarily in the gas for $\alpha$ rays and for low energy $\beta$ or $\gamma$ (X) rays; high-energy incident radiations cause the liberation of secondary electrons in the walls, which, in turn, cause the formation of ion pairs in the gas. Ionization chambers and counters for the former radiation should, therefore, be provided with thin windows or thin walls, while the wall thickness for the more energetic radiation should be made "greater than the maximum range of the secondary electrons, yet not so thick as to produce any appreciable attenuation of the primary gamma radiation."[1]

A multiple-plate ionization chamber for X rays is shown in Fig. (5-2)6. The chamber has the advantage that its volume and, therefore, its sensitivity are large, yet the spacing between the electrodes is small and the time required for the collection of electrons and ions is short. Therefore, the resolving time for very fast X-ray pulses is short (on the order of a fraction of a millisecond).

A chamber type frequently used for X-ray dose determination is illustrated in Fig. (5-2)7. Chambers of this kind can be built quite small (less than 1 cm$^3$) for application in body cavities or for "point-by-point" measurement of dose distribution.

A chamber primarily useful for the measurement of $\alpha$ rays is shown in Fig. (5-2)8.

Extensive descriptions of ionization chambers for use in nuclear research will be found in Rossi and Staub, *loc. cit.* Ionization chambers for use in X-ray dosimetry are reviewed by R. Jaeger, *Arch. tech. Messen,* V 61–1, January, 1941. For further descriptions and extensive literature, see S. A. Korff, "Electron and Nuclear Counters," D. Van Nostrand Company, Inc., Princeton, N.J., 1946, and D. H. Wilkinson, "Ionization Chambers and Counters," Cambridge University Press, London, 1950.

## 5-22. GAS-FILLED COUNTERS

*a. Geiger Counters with Cylindrical Geometry.* The most common form of counter is shown schematically in Fig. (5-2)9. The anode consists of a thin tungsten or platinum wire held by insulators within a cylindrical cathode. The cathode is thin enough, or has a window, to permit radiation to enter the tube. The counter is filled with a gas, such as argon, to which is added a small amount of polyatomic gas (e.g., alcohol). The pressure in the counter is in general between a few centimeters Hg and atmospheric pressure. A voltage is applied through a resistance $R$ and causes an electric field in the counter which is strongest around the anode. If an ionizing particle enters the counter and causes ionization, a discontinuous electric gaseous

[1] B. B. Rossi and H. H. Staub, "Ionization Chambers and Counters," McGraw-Hill Book Company, Inc., New York, 1949.

discharge will take place in the counter. As the result of this discharge, an intermittent current will flow and a voltage pulse will appear across the resistance $R$. The current can be many orders of magnitude larger (up to $10^9$) than the current caused by the movement of the electrons and ions formed by the ionizing particle.

Four different modes of operation of the counter can be distinguished, depending primarily upon the magnitude of the voltage applied to the counter:

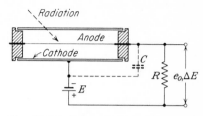

FIG. (5-2)9. Geiger–Müller counter and basic circuit.

REGION A: IONIZATION CHAMBER REGION, FIG. (5-2)10. An ionizing particle, such as an $\alpha$ or $\beta$ particle, entering the tube will produce electrons and ions. If a voltage is applied, the charges will be separated; the electrons will drift toward the anode, and the positive ions toward the cathode. If the voltage applied is high enough to move the electrons and ions toward the electrodes before an appreciable fraction of them recombine, yet not so high that ionization by collision occurs, the number of charges moving toward and arriving at the electrodes is equal to the number produced by the initial ionizing event. The migration of the electrons and ions toward the electrodes causes a current and the appearance of an output signal across the resistor $R$, which is proportional to the number of ions $n$ formed by the incident ionizing radiation:

$$\Delta E = kn$$

The counter acts as an ionization chamber. The size of the pulse

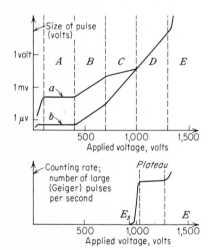

FIG. (5-2)10 (*top*). Operating characteristics of counters: $A$, region of ionization chamber; $B$, region of proportionality; $C$, region of limited proportionality; $D$, Geiger region; $E$, region of unstable operation.

FIG. (5-2)11 (*bottom*). Counting rate versus applied voltage of a Geiger counter.

is independent of the applied voltage within the region $A$ (see Ionization Chambers, 5-21).

REGION B: PROPORTIONALITY REGION. As the applied voltage is raised the kinetic energy of the electrons and ions increases, so that between collisions they will accumulate enough energy to create new electrons and ions by inelastic impact with neutral molecules. This process arises first in the close vicinity of the wire (anode) where the field strength is highest. Each electron and ion formed by collision may produce new ions; the current increases in the form of an avalanche (Townsend avalanche). If $n$ electron-ion pairs are formed by the incident particle and if $(A - 1)$ ion pairs are formed by each initial electron on its way toward the anode, then $A \times n$ ion pairs will migrate toward and be collected by the electrodes, and the output signal will be

$$\Delta E = kAn$$

The size of the output pulse is, therefore, proportional to $n$, the number of ions initially formed by the incident radiation. $A$ is the amplification factor; its value depends upon the applied voltage as well as upon the wire radius, the capacitance of the counter system, the gas pressure, and the kind of gas. The value of $A$ is between about 1 and 100 for proportional counters filled with monatomic and diatomic gases and can rise to $10^4$ for counters filled with polyatomic gases. The discharge in the counter ceases when all electrons and ions are collected at the electrodes.

The magnitude of the proportional pulse can be as high as 10 to 100 mV. The pulse rises to a maximum value in a fraction of a microsecond; the decay time depends upon the time constant of the network following the counter. In general, the entire counting process has a duration of several microseconds. After that time the counter is ready for the recording of another pulse.

The proportional counter permits the measurement of the number of particles caused by primary ionization and, therefore, of the energy of the ionizing particles. It is, furthermore, useful for the observation of highly ionizing particles, such as $\alpha$ particles, in the presence of a low ionizing radiation (e.g., $\gamma$-ray background).

REGION C: REGION OF LIMITED PROPORTIONALITY. As the applied voltage is further raised, the behavior of the counter changes primarily in two respects: (1) For low-energy ionizing particles, the size of the output pulse increases with increasing voltage faster than in the proportional region, and (2) for high-energy ionizing particles, the pulse size rises with increased applied voltage, but to a lesser degree. Therefore, in the region $C$, the output pulse is not proportional any more to the initial ionization (number of ion pairs formed by the ionizing particle), until at the upper limit of the region of limited proportionality all pulses have the same size regardless of whether they are initiated by one or many ion pairs. The amplification factor can rise as high as $10^7$. Counters are rarely used in the region of limited proportionality; the description of the physical mechanism is omitted, therefore.

REGION D. GEIGER-COUNTER REGION. The characteristic property of a counter operated in the Geiger region is the appearance of output pulses that are all of the same size, independent of the number of ions initially formed by an incident ionizing particle or radiation.

The physical mechanism which leads to the formation of a Geiger pulse is initially the same as that giving rise to a proportional pulse: An incident ionizing particle produces an electron-ion pair. The electron drifts toward the positive central wire and, within a narrow region around it, causes ionization by collision and the formation of an avalanche.

Before the electrons and ions formed in the avalanche are collected at the electrodes, a second phenomenon takes place. Light emitted from excited atoms in the first avalanche causes photoionization of the gas molecules and the liberation of new electrons which cause further avalanches. The light emission proceeds in every direction, but only in the vicinity of the central wire is the field strength high enough to facilitate the formation of new avalanches. The discharge spreads along the wire, therefore, until the entire anode is surrounded by a narrow cylinder of ions, and the discharge becomes self-sustaining.

The termination of the discharge comes about in the following manner. The positive ions surrounding the central anode move toward the cathode with a velocity which is considerably smaller (because of their mass) than that of the electrons. These ions cause a positive space charge surrounding the positive wire; as they move toward the cathode, the positive space-charge cylinder ("the virtual anode") increases in diameter and, therefore, the field strength in the counter decreases. [The field strength at a radius $r$ between the cylinder radius $r_c$ and the wire radius $r_a$ is $F = E/r \log (r_c/r_a)$.] The discharge ceases when the field strength within the counter has fallen below the level where a discharge can be sustained.

The pulse size of Geiger counters is, in general, between 1 and 10 volts and sometimes higher; the amplification factor is between $10^7$ and $10^9$.

The discharge will be reignited if the positive ions arrive at the cathode and liberate new electrons. To avoid this possibility, three different methods of "quenching" the counter are available.

1. Resistance quenching. The capacitance of the counter is expressed by the equivalent capacitor $C$ in Fig. (5-2)9. The gas-discharge process reduces the voltage on this capacitor; the capacitor is recharged through the resistor $R$. If the time constant $CR$ is made longer than the time required to sweep the ions out of the counter, no reignition will take place. The method requires time constants of the order of $10^{-2}$ sec; during this time the counter is inoperative, and the method cannot be used for fast counting.

2. Electronic-quenching methods. A number of systems have been devised in which the output pulses from the counter cause an electronic circuit to reduce or remove the voltage on the counter temporarily and then to recharge the counter. The systems can be made to operate in about $10^{-4}$ sec.[1]

3. Self-quenching counters. The admixture to the inert gas in the counter of a small amount (about 10 per cent) of a polyatomic gas (i.e., generally a gas with four and more atoms, frequently a hydrocarbon such as methane or alcohol vapor) changes the quenching characteristic of the counter profoundly and prevents a reignition of the discharge. The photons emitted from the monatomic or diatomic gas interact with the polyatomic gas. Thus, these photons are prevented from reaching the cathode where they may liberate new electrons. Instead, they are absorbed by the heavy organic molecule, which is either predissociated or decomposed. If the energy is finally carried to the cathode surface, it is too small to cause a photoelectric emission. Counters filled with such gases usually have a higher starting potential than non-self-quenching counters, their lifetime is shorter because of the decomposition of the quenching agent in each discharge, and they show variation of their behavior at low temperatures, where condensation of the quenching agent may occur.

If the counter is exposed to ionizing radiation and the voltage across the counter is raised, the counting rate will first increase, as in Fig. (5-2)11. Then follows a region, the plateau, where the counting rate will only slightly increase. This is the region in which the counter should be operated. In the plateau region the number of discharges is directly related to the number of ionizing particles striking the tube and does not depend appreciably upon the applied potential. No counter has an ideal plateau of a slope zero; at best

[1] See Korff, *op. cit.*, chap. 7B.

the plateau rises with <1 per cent/100 volts. At the end of the plateau the counting rate increases again, and the counter may produce self-sustained, continuously repeating pulse discharges even in the absence of external ionizing radiation. Operation of the counter in this region, Fig. (5-2)10$E$, is likely to damage the counter.

Figure (5-2)11 is not to be interpreted in the sense that the number of counts below the starting potential $E_s$ is zero. Rather, a large number of small pulses can be observed at voltages below $E_s$, proportional pulses of different sizes, but smaller than Geiger pulses. The number of proportional pulses counted depends upon the sensitivity (discriminator setting) of the subsequent amplifier stage.

If an electron or ion pair is produced by incident ionizing radiation, the counter will start to respond after a very short time delay $t_0$, as shown in Fig. (5-2)12. This time delay is due to the electron transit time; it is of the order of a fraction of a microsecond and may exceed 1 $\mu$sec only in large counters. The output pulse rises to its maximum value in a few microseconds and decays within several hundred microseconds. During this time the counter does not register any ionization caused

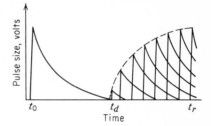

Fig. (5-2)12. Pulse height–time diagram of a Geiger counter.

by a new incident particle or radiation (dead time). After that time there follows a period (recovery time) where the counter will respond with small pulses; toward the end of the recovery time the pulse increases to normal size.[1] Also, the recovery time is of the order of several hundred microseconds. It is shorter in non-self-quenching counters than in those filled with a quenching agent.

The highest level of radiation for which a Geiger counter is suitable is determined by the recovery time of the counter. In commercial counters, this time is of the order of 30 to 300 $\mu$sec; the highest counting rate is, therefore, $3 \times 10^3$ to $3 \times 10^4$ cps.

In the plateau region, the number of counts is proportional to the number of primary ionizations provided the ionizing particles arrive at a sufficiently low rate. As the number of incident ionizing particles or the incident radiation rises, the counter cannot follow any longer; the number of counts, as well as the average output current, reaches a maximum and then declines with further increase of incident radiation. The situation can be remedied, and the region of

[1] H. G. Stever, *Phys. Rev.*, **61**, 38 (1942).

incident radiation for which the counter is suitable can be extended by operating the counter intermittently, i.e., with square pulses. A discharge will then occur only if an ionizing particle traverses the counter during the time when the square pulse is applied to the counter.[1] The scale factor, i.e., the ratio of counting rate for continuous (d-c) operation to that of intermittent (pulse) operation, is $F_s = 1/wf$, where $w$ is the pulse width and $f$ the pulse repetition rate. Besides an extension of the useful range of the counter toward higher levels of radiation, the pulsed operation offers two further advantages: (1) The rate of spontaneous counts is reduced (Rossi

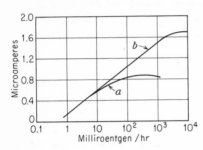

and Staub describe a counter of 0.125 in. diameter which is pulsed six times per second and furnishes not more than six spontaneous counts per hour), and (2) the size of the pulse is greatly increased (on the order of 1,000 volts) because, for the duration of the pulse, the voltage can be raised above the sparking potential.

Fig. (5-2)13. Characteristics (average current versus incident radiation) of a Geiger counter: (a) for d-c operation; (b) for pulse operation [*from S. W. Lichtman, Nucleonics, 11, 22 (1953); by permission*].

Two curves representing the average counter current versus radiation intensity for d-c and pulsed operation are shown in Fig. (5-2)13. The linear range of the average tube current versus incident radiation intensity is increased without altering the performance of the tube at lower radiation levels.

The lowest level of radiation that can be measured with a Geiger counter depends upon the background counting rate and upon the error that can be tolerated. If the number of background counts in any time interval $t$ is $C_B$, the number of counts resulting from incident radiation (the radiation to be measured) is $C_s$, and if $C_T$ ($= C_s + C_B$) is the number of total counts measured in the time interval $t$, the statistical probable error of $C_T$ is

$$\epsilon = \frac{\sqrt{C_T + C_B}}{C_s} = \frac{\sqrt{C_s + 2C_B}}{C_s}$$

If $S$ is the rate at which ionizing particles of the radiation to be measured enter the counter, $\eta$ the counter efficiency, and $B$ the

---

[1] B. B. Rossi and H. H. Staub, *op. cit.*, p. 117; S. W. Lichtman, *Nucleonics*, **11**, 22 (January, 1953).

background counting rate, the probable error can be expressed by

$$\epsilon = \frac{\sqrt{\eta S + 2B}}{\eta S} \frac{1}{\sqrt{t}}$$

The error, at a given level of radiation, diminishes in proportion to the root of the observation time $t$. The minimum radiation intensity $S$ that can be measured with a given error, or the time required to obtain a measurement, diminishes, of course, with a reduction of the background counting rate. An increase of the counter efficiency $\eta$ does not necessarily produce the same result, since it may also lead to an increase of the background counting rate.[1] The background counting rate of commercial counters is between 1 and 50 counts/min; the efficiency, for instance for the $\beta$ radiation of $C^{14}$,

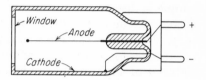

FIG. (5-2)14. Cross section of a Geiger-counter tube with window.

is between 2 and 10 per cent.

Commercial Geiger counters are predominantly of two types, the cylindrical type as illustrated in Fig. (5-2)9 and the end-on type as shown in Fig. (5-2)14.

The cylindrical type is primarily used for gamma radiation and, if the walls are sufficiently thin, for the more energetic $\beta$ radiation. The wall thickness of commercial counters expressed in absorbing mass per unit wall surface area can be made as small as 30 mg/cm², but is in general above 300 mg/cm². Counters vary in length from 2 to 40 in. and in diameter from 0.125 to 2 in. The operating voltage is between 400 and 1,500 volts; the length of the plateau is around 200 volts, the slope of the plateau of the order of 1 to 10 per cent/100 volts. The recovery time of the counters varies from less than 100 to 400 μsec.

End-on-type counters are used in general for the counting of α and $\beta$ rays and soft X rays which can penetrate the thin windows usually made from mica or Mylar (trade name, E. I. du Pont de Nemours & Company). Counters of the type shown in Fig. (5-2)14 can also be used open (windowless counter). The radioactive specimen is placed in front of the opening within a closed housing, and the filling gas (for Geiger counters, 99 per cent helium plus 1 per cent isobutane; for proportional counters 90 per cent argon plus 10 per

---

[1] See *Tracerlog*, no. 84, p. 7, issued by the Tracerlab, Inc., Boston, April, 1957.

cent methane) flows in a slow stream through the counter. The relative yield of these counters for the soft $\beta$ radiation of carbon 14 (0.155 MeV) is approximately

| Counter | $n$, % |
|---|---|
| Windowless | 35–45 |
| Mylar window (0.9 mg/cm$^2$) | 22–25 |
| Mica window (2 mg/cm$^2$) | 10 |

A well-type counter is illustrated in Fig. (5-2)15. The radioactive sample is inserted in the center; the counter is, in effect, a multiplicity of counters, which surround the sample so that radiation emitted from the specimen in almost every direction is counted.

Cathodes for gamma-ray counters should preferably be made of a material of high atomic number $Z$, since the number of photoelectrons produced by incident gamma radiation increases with $Z$. For instance, counters with bismuth cathodes ($Z = 83$) have a higher counting efficiency for gamma rays than those with copper cathodes ($Z = 29$). The superiority of the bismuth counter depends upon the energy of the incident quanta; at low gamma-ray energies, the bismuth counter can be superior by a factor of 5 or more;[1] at gamma-ray energies beyond about 1 MeV, the superiority of the high-$Z$ cathode counter is only of the order of ten to thirty per cent.[2]

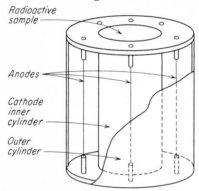

Radioactive sample

Anodes

Cathode inner cylinder

Outer cylinder

FIG. (5-2)15. Well-type counter.

Cathodes in counters filled with chlorine are made with titanium and zirconium.[3] Counters with a semiconducting cathode of tin oxide coating fused on glass have been described by Clark.[4] The thin tin oxide layer, which is transparent, is not attacked by chlorine. Counters of this type can be made from thin wall glass of about 3 mg/cm$^2$.[5]

[1] Rossi and Staub, *op. cit.*, chap. 5.2.

[2] R. D. Evans and R. A. Mugele, *Rev. Sci. Instr.*, **7**, 441 (1936); also H. Maier-Leibnitz, *Z. Naturforsch.*, **1**, 243 (1946).

[3] A. L. Ward and A. D. Krumbein, *Rev. Sci. Instr.*, **26**, 341 (1955).

[4] L. B. Clark, *Rev. Sci. Instr.*, **24**, 641 (1953).

[5] *Bulletin*, Radiation Counter Laboratories, Skokie, Ill.

All-glass counters have been described by Maze.[1] Connection is made by means of a layer of colloidal graphite ("Aquadag") applied to the outside of the glass envelope. The effective resistance of the glass wall is of the order of $10^8$ ohms.

The sensitivity to gamma rays can be increased by increasing the surface area of the cathode, for instance, by employing screen or metal gauze cathodes. The gain is moderate; an increase of the surface area by a factor of 4 increases the relative sensitivity by about 50 per cent.

The lifetime of a self-quenching counter is limited by decomposition of the quenching agent. Good counters have a life of the order of $10^{10}$ counts; counters filled with methane as a quenching agent have a lifetime of about $10^8$ counts. At the end of its useful life the operating potential of the counter, the slope of the plateau, and the number of spurious discharges increase, and multiple pulses ("bursts") begin to appear. Operation of the counter near the lower limit of the plateau is recom-

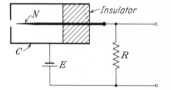

Fig. (5-2)16. Geiger point counter.

mended to increase the lifetime. Because of the absence of decomposition of the quenching agent, the lifetime of non-self-quenching counters can be considerably longer.

Design considerations and constructional details of counters for alpha-, beta-, gamma-, and X-ray counters, as well as detectors for neutron recoils $(n,\alpha)$ and $(n,p)$ reactions and fission detectors, may be found in Rossi and Staub, *op. cit.*, chaps. 5–9. For further references, see D. H. Wilkinson, *loc. cit.*; S. A. Korff, *op. cit.*; H. Friedman, *Proc. IRE*, **37**, 791 (1949); D. R. Corson and R. R. Wilson, *Rev. Sci. Instr.*, **19**, 207 (1948); G. J. Hine and G. L. Brownell, "Radiation Dosimetry," Academic Press, Inc., New York, 1956; J. Sharpe, "Nuclear Radiation Detectors," Methuen & Co., Ltd., London, and John Wiley & Sons, Inc., New York, 1955.

*b. Point Counters.* The counter type originally used by Geiger is the point counter illustrated in Fig. (5-2)16. It consists of a needle $N$ held by an insulator within a cylindrical metal chamber $C$. An open window in front of the needle point permits ionizing radiation to enter the counter. The point counter is generally used in air at atmospheric pressure; it can be used with the needle at a positive or at a negative potential; the discharge mechanism and the behavior of the counter are, of course, different with different polarity.

[1] R. Maze, *J. phys. radium*, **7**, 164 (1946).
[2] Evans and Mugele, *loc. cit.*

Point counters filled with argon or nitrogen and a quenching agent have been investigated by Lion and Vanderschmidt.[1] The counting rate increases first sharply with the applied voltage, as shown in Fig. (5-2)17, region $A$. With further increasing voltage there follows a region ($B$) of stable counting in which the counting rate increases linearly with the voltage. At still higher voltages the counting rate increases again rapidly ($C$), and the operation becomes unstable. The quenching agent has the effect of broadening the region $B$ as

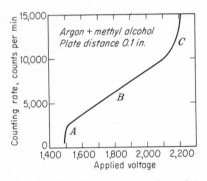

FIG. (5-2)17. Characteristic of a point counter. Counting rate as function of the applied voltage (atmospheric pressure, argon + 12 per cent alcohol, point-to-plane distance 0.1 in. [*from K. S. Lion and G. F. Vanderschmidt, J. Opt. Soc. Am.*, **45**, 1024 (1955); *by permission*].

FIG. (5-2)18. Parallel-plate counter.

well as of shifting it to higher voltages. A plateau like that in cylindrical counters, Fig. (5-2)11, does not exist. The critical volume (volume in which an electron must enter to trigger the discharge) is considerably smaller than that of the cylindrical counter and is commonly assumed to be located in a cone in front of the needle.

*c. Parallel-plate Counters.* The counter is shown schematically in Fig. (5-2)18. The distance between the electrodes is usually a fraction of a millimeter. One of the electrodes must be thin enough to permit ionizing particles to enter the counter. The discharge mechanism is initiated by the liberation of an electron from the counter wall (only rarely by gas ionization). The electron forms an avalanche and, at low voltages, an output pulse of the order of millivolts. As the voltage is raised, spark discharges occur which can be observed visually through a transparent electrode. The output pulses are of

[1] K. S. Lion and G. F. Vanderschmidt, *J. Opt. Soc. Am.*, **45**, 1024 (1955).

the order of several hundred volts.  The counting rate versus voltage characteristic is similar to that shown in Fig. (5-2)17.  For stable counter operation a series resistor greater than $10^6$ ohms must be used, as well as a quenching agent.[1]

The discharge in the parallel-plate counter does not spread throughout the entire counter but remains localized in that region where the incident particle causes the liberation of a primary electron.  The discharge is terminated when the voltage across the counter has fallen below the minimum sustaining value.  The capacitance of the parallel-plate capacitor is larger than that of the cylindrical or the point counter;  the dead time and the recovery time are

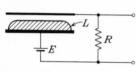

FIG. (5-2)19.  Parallel-plate counter with resistive layer *L*.

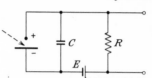

FIG. (5-2)20.  Spark counter, schematic diagram.

large, therefore, and incident ionizing particles that arrive during these times at any point of the counter are not recorded.  Lion and Vanderschmidt[2] have overcome this disadvantage by placing a resistive layer *L* (ferrite) in the counter, as shown in Fig. (5-2)19.

*d. Spark Counters.*  The discharge in a non-self-quenching counter can be terminated by a series resistance which causes a reduction of the voltage across the counter whenever a discharge occurs (see Region D, Geiger-counter Region, 5-22a).  If a capacitor *C* is connected in parallel to the discharge gap, as shown in Fig. (5-2)20, the charge stored in it causes the gas discharge to assume the character of a spark that is strong enough to be seen or photographed or heard without further amplification (spark counter).[3]  Of course, the large time constant of the counter, imposed by the resistance *R* and the capacitance *C*, reduces the time resolution of the counter; yet the system is simple and useful for a number of applications, e.g., for X-ray dosimetry.  The discharge in the Greinacher counter takes place in a point-to-plane gap or between two small spheres operated in air at atmospheric pressure.  The counter characteristic does not exhibit a plateau.

[1] R. W. Pidd and L. Madansky, *Phys. Rev.*, **75**, 1175 (1949).

[2] Lion and Vanderschmidt, *loc. cit.*

[3] H. Greinacher, *Helv. Physica Acta*, **7**, 360 and 514 (1934);  other references by P. Frey, *Helv. Physica Acta*, **19**, 41 (1946).

A similar counter system described by Chang and Rosenblum[1] and illustrated in Fig. (5-2)21 consists of a wire stretched tautly parallel to a metal plane. The plane has rounded edges to reduce the field strength and to prevent flashover at these ends. The counter has a plateau region between 4,500 and 6,500 volts. The rate of spurious discharges is less than 1 count in 4 hr, and the dead time is about $10^{-4}$ sec.[2]

### 5-23. Solid-state Transducers for Ionizing Radiation

*a. Crystal Counters.* The crystal counter, schematically illustrated in Fig. (5-2)22, consists of a slab of a crystal such as silver chloride

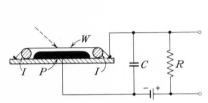

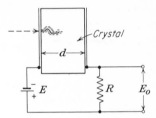

FIG. (5-2)21. Spark counter with wire electrode $W$ held in a position parallel to a plate $P$ by means of insulators $I$.

FIG. (5-2)22. Crystal counter, schematic diagram.

(AgCl) or diamond (C) on which thin metal electrodes are deposited. A voltage source $E$ causes a field strength of the order of several kilovolts per centimeter in the crystal. If ionizing radiation, for instance an energetic beta particle, enters the crystal, it will cause free electrons in the vicinity of its path, analogous to ionization in a gas. Accelerated by the applied field, the electrons will move through the crystal until they are trapped or reach the positive electrode, thus forming a current and a voltage pulse. The magnitude of the pulse is of the order of 1 mV. If an ionizing particle enters the crystal and causes the release of $n_0$ electrons in the vicinity of the negative electrode, the migration of the released electrons gives rise to a voltage pulse of the magnitude

$$E_o = \frac{n_0 e}{C} \frac{\lambda}{d}(1 - e^{-d/\lambda}) \tag{1}$$

where $e$ is the electron charge, $C$ the capacitance of the crystal plus that of the associated circuit, and $d$ the crystal thickness. The

[1] W. Y. Chang and S. Rosenblum, *Phys. Rev.*, **67**, 222 (1945).
[2] R. M. Payne, *J. Sci. Instr.*, **26**, 321 (1949).

magnitude $\lambda$, the "Schubweg," is the average distance moved by a cluster of electrons in the crystal; it depends upon the crystal material and is proportional to the applied field strength. The pulse size rises first linearly with the applied voltage and approaches saturation at higher field strengths. In the linear part the pulse size is directly proportional to the energy of the incident particle. If the incident particle penetrates deeply into the crystal and causes the release of electrons uniformly throughout the crystal volume, Eq. (1) changes to

$$E_o = \frac{n_0 e}{C} \frac{\lambda}{d}\left[1 - \frac{\lambda}{d}\left(1 - e^{-d/\lambda}\right)\right]$$

A crystal acts as a good counter only if the Schubweg $\lambda$ has a magnitude from a few tenths of a millimeter to 1 cm. The number of electrons released by an incident particle of a given energy depends upon the energy required per ion pair formation. In a crystal, this energy is about 5 to 10 eV as compared to about 30 eV in a gas. Only a fraction (e.g., 60 per cent) of the energy of the incident particle is used to release electrons; a 1-MeV beta particle causes in a silver chloride crystal about $1.2 \times 10^5$ electrons. The lowest detectable energy for an incident particle is of the order of 1 keV (100 to 200 electrons).

A polarization effect takes place in some crystals. The immobile holes in some crystals form a positive space charge near the negative electrodes; the electrons trapped near the positive electrode form a negative space charge. These space charges eventually reduce the effect of the applied field and the size of the output pulses. Under certain conditions, the polarization effect can be overcome by periodically reversing the applied field. Another possibility to reduce the positive space charges is to irradiate the crystal with infrared light. The polarization effect limits the lifetime of the crystal counter. If the polarization cannot be removed, a counter exposed to an incident gamma radiation of 1 MeV will have a lifetime of $10^6$ counts/cm$^3$ of crystal material; exposed to an alpha radiation of 5 MeV energy, it will have a lifetime of $10^8$ counts/cm$^3$. The lifetime is practically unlimited when the space charges are removed.

The output pulse and, therefore, the resolving time of the counter are considerably shorter than that of the ionization chamber or of the Geiger counter. The resolving time depends upon the applied field; for a field strength of about 5 kV/cm the resolving time of the silver chloride or silver bromide counter is about 0.2 $\mu$sec, that of the diamond counter only 0.04 $\mu$sec.

A number of materials have been found suitable for counting, the most frequently used crystals are silver chloride, silver bromide, and thallium bromide–thallium iodide at the temperature of liquid air, and diamond, sulfur, and cadmium sulfide at room temperature.

Two groups of cadmium sulfide crystals exist, a nonluminescent group and an impurity-activated luminescent group. The former have normal counting properties, the pulse size being determined by the number of electrons released by the incident particle. The luminescent cadmium sulfide crystal exhibits a multiplication effect; the incident ionizing particle causes the release of electrons which give rise to a semipermanent conduction phenomenon in the counter and, in further sequence, to a voltage pulse that is enormously larger than in the nonluminescent crystal. A single alpha particle may cause a pulse of 10 volts; the duration of this pulse is of the order of 0.1 sec, several orders of magnitude larger than that of the ordinary (silver chloride) crystal counter.

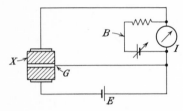

Fig. (5-2)23. Arrangement for the measurement of radiation-induced photoconductivity: $X$, crystal; $E$, voltage source (300 volts); $G$, guard ring to eliminate the effect of surface current; $B$, compensation arrangement to eliminate the effect of dark current.

Many materials require careful preparation before they count, and not all specimens of the same material show any counting action. Annealing and careful heat treatment are required for silver and thallium halides. Suitable measures have to be taken to avoid strain; gas ionization in the vicinity of the crystal can be avoided by evacuation. Adequate insulation and low-capacitance mounting must be provided. The amplifier following the counter must have a fast rise time and a low noise level. The crystal should be kept in darkness to avoid photoconduction.

Extensive lists of references may be found in the review papers by R. Hofstadter, *Nucleonics*, **4**, 2 (April, 1949), and 29 (May, 1949); *Proc. IRE*, **38**, 721 (1950).

*b. Radiation-induced Photoconductivity.* Alkali halide crystals that are exposed to high-energy radiation (e.g., gamma or X rays) become colored and in this state are photoconductive. If such crystals are afterwards exposed to light, they will be bleached again. If a voltage is applied to the crystal, as shown in Fig. (5-2)23, a current will flow during the bleaching process. The magnitude of the current pulse is proportional to the X-ray dose to which the crystal was exposed.

The method has been investigated with potassium bromide crystals by Lichtenstein.[1] The magnitude of the output pulse is proportional to the X-ray dosage in the range between 2.5 to about 5,000 mr. The lowest value of radiation dose to be measured is imposed by the residual (dark) photoconductivity of the crystal and is about $\frac{1}{4}$ mr. A dosage of 1 roentgen given to a crystal of 1 cm$^3$ results in an output pulse of about 5 volts. The sensitivity varies for different crystals of the same type by about 1 to 3; the variation is probably due to differences in impurities.

[1] R. M. Lichtenstein, AIEE-IRE Conference on Electronic Instrumentation in Nucleonics and Medicine, New York, Oct. 31–Nov. 2, 1949.

# Name Index

Adams, G. D., 202
Aiken, C. B., 278
Alexander, A. L., 14
Allegheny Ludlum Steel Corp., 200
Allen, C. M., 118
Allen, J. S., 215
Allen, P. J., 213
Alpert, D., 113
Amdur, I., 278
American Institute of Physics, 154
Andrew, E. R., 206, 208
Andrews, D. H., 279
Anrep, G. V., 118
Arnold, F. S., 122
Arons, A. B., 105, 106
Astin, A. V., 141
Awbery, J. H., 136

Babbit, J. D., 157
Bahr, G. F., 216
Bailey, A., 130
Baker, E. B., 278
Bancroft, D., 105
Barber, W. C., 22
Barnes, G. W., 262
Barr, E. E., 278, 280
Bates, L. F., 178
Bauer, G., 279

Bauer, H., 262
Baumberger, R., 35
Bay, Z., 215
Bayard, R. T., 113
Beattie, J. A., 158
Becker, A., 216
Becker, J. A., 161, 198
Becker, M., 272
Beckman Company, 145
Bennett, A. I., 257, 258
Bennett, R. S., 12
Benzer, S., 268
Berger, F. B., 28, 62
Bergmann, L., 17
Bernard, R., 182
Betz, P., 232
Beynon, J. H., 115
Bhalla, M. S., 113
Billings, B. H., 278, 280
Birnbaum, G., 146
Blacet, F. E., 276
Blackburn, John F., 28, 61, 72, 94
Bloch, F., 208
Bloembergen, N., 207
Böhme, W., 17
Boiten, R. G., 36
Boucke, H., 199
Bouwcamp, C. F., 135
Bowse, H. A., 99

305

# Subject Index

Acceleration transducers, 86, 91
Air conductivity, 220
Air-equivalent chamber, 286
Alpha gauge, 22–23
Alpha-ray-sensitive transducers,
  alpha-ray Geiger counter, 287
  based on gas ionization, 281–300
  photoconductive, 264, 266–267
  solid-state systems, 300–303
Ammonium dihydrogen phosphate,
  piezoelectric properties, 80
Atmospheric electricity transducers,
  218–220, 223–225


Ballisto-cardiograph, 63
Barium titanate ceramics, piezo-
  electric properties, 80
Beckman Hygrometer, 145
Becquerel effect, 170
Beta gauge, 21–22
Beta-ray-sensitive transducers, based
  on gas ionization, 281–300
  solid-state systems, 300–303
Beta rays (*see* entries under Electron)
Bimorph crystals (*see* Piezoelectric-
  displacement transducers)
Bismuth spiral for magnetic field
  measurements, 196–199

Blackening of thermal-radiation
  transducers, 276–277
Bolometers, for displacement
  measurements, 72–74
  for radiation measurements,
  278–280


Cables, noise originating in, 71
Callendar equation, 158
Capacitive transducers for measurement of displacement, 66–72
  humidity, 146–150
  liquid level, 98–100
  pressure, 104–105
  surface charge or surface potential,
  227–229
  temperature, 166–167
  thickness, 14–15
Capacitors, 14–15, 66–72
  differential, 67–70
  fringe effect, 14–15, 66
  serrated, 70
  variable, 66–70
    change-of-area system, 69–70
    change-of-distance system, 66–69
Cathodes in photoemissive systems,
  235–237
Charged-plate ion collector, 216